T0141392

SEEING UNDERGROUND

SEEING

UNDERGROUND

MAPS, MODELS, AND MINING ENGINEERING IN AMERICA

Eric C. Nystrom

University of Nevada Press
RENO & LAS VEGAS

This work was made possible in part by the
Paul A. and Francena L. Miller
Research Fellowship, College of Liberal Arts,
Rochester Institute of Technology.

University of Nevada Press, Reno, Nevada 89557 USA
Copyright © 2014 by University of Nevada Press
All rights reserved
Manufactured in the United States of America
Design by Kathleen Szawiola

Library of Congress Cataloging-in-Publication Data
Nystrom, Eric Charles.
Seeing underground : maps, models, and mining engineering in
America / Eric C. Nystrom.
pages cm
Includes bibliographical references and index.
ISBN 978-0-87417-932-3 (cloth : alk. paper) —
ISBN 978-0-87417-933-0 (e-book)
1. Mine maps—United States—History. 2. Mines and mineral
resources—United States--History. 3. Geological mapping—
United States—History. I. Title.
TN273.N97 2014
622'.140973—dc23   2013039139

The paper used in this book meets the requirements of American
National Standard for Information Sciences—Permanence of Paper
for Printed Library Materials, ANSI/NISO Z39.48-1992 (R2002).
Binding materials were selected for strength and durability.

First Printing
23 22 21 20 19 18 17 16 15 14
5 4 3 2 1

CONTENTS

ILLUSTRATIONS

PREFACE

I hadn't planned to study mining history, but the first time I saw an underground map, I was thunderstruck. The vivid colors, the tangle of angular lines, and the lack of background decoration suggested free jazz or abstract art, not an engineering document. I couldn't figure out what it was really *for*—what question could this thing possibly be the answer to? The disconnect between the map's weird beauty and its inscrutable utility captivated me as I searched for explanations, but few were forthcoming. A breakthrough occurred when I realized, after a long day of examining maps from a coal mine, that my hands weren't *dirty enough*. The maps carried ordinary archival grime, but were too clean to have ever been underground. But if they weren't used underground, I reasoned, they couldn't have been used by the miners to find their way around down below. Instead, I realized, the maps must have lived in an office—in the engineer's office. Why would engineers make maps that weren't used in the places they represented? Inspired, I continued my search for answers, with new attention to the makers of these maps—mining engineers.

I realized in time that there was a bigger story here, a story about the emergence of a technical profession and the tools those professionals used to conduct their work. Experienced miners in the nineteenth century had little use for underground maps. The proliferation of maps was closely associated with the mining engineers who made and used them. In fact, underground maps (and their three-dimensional equivalent models), together with the practices, discourses, and material objects that were associated with their creation and use, formed a distinctive visual culture of mining engineering. Making and using maps and models to understand and attempt to control underground mines became a key part of what it meant to be a mining engineer, and set these engineers apart from experienced miners. Their professional training involved gaining facility with the visual culture of mining; as the profession rapidly evolved, so too did the maps and models. Indeed, the ability to wield the visual culture of mining helped make the case for why mining engineers, and not experienced miners, should be preferred for the management of complex underground mines, a struggle that was all but won by the early decades of the twentieth century.

Perhaps ironically, underground mines are difficult places in which to see, and visibility has long been a key ingredient of authority and control. Mines are profoundly dark, and a visitor's lamp appears as only a suggestion of light against the blackness. These mines cannot be seen from the surface either, except for their hoisting and processing works, which give little sense of the vast tunnels and shafts radiating below. Only the visual culture of mining engineering can make them visible at a glance—a powerful trick indeed.

An enormous number of people had a hand in shaping this project, from initial idea to the book in your hand.

Bill Leslie and the late Hal Rothman were extraordinary advisers, challenging and encouraging me, helping open doors, and serving as role models, and I am enormously grateful to them. Peter Liebhold and Steve Lubar were instrumental in shaping my thinking about material culture in particular, as well as the history of technology, and that fresh way of looking at the sources had an important impact on the shape of my scholarship.

My colleagues and administrators at the Rochester Institute of Technology (RIT) have provided a supportive environment and, on several occasions, additional resources. The Paul A. and Francena Miller Fellowship, awarded by the College of Liberal Arts at RIT, supported final revisions, image permissions, and indexing.

Historians who are interested in mining have helped me wrestle with these concepts and encouraged my work, especially Roger Burt, Richard Francaviglia, Chris Huggard, Ron James, Brian Leech, Catherine Mills, Jeremy Mouat, Fred Quivik, Terry Reynolds, Duane Smith, Bob Spude, and David Wolff. So many members of the Mining History Association have been terrific—generous with their knowledge and quick to share examples of maps and models—that the book is much richer as a result. Thank you to the late James Bohning, and to Eric Clements, Terry Humble, Johnny Johnsson, Mike Kaas, Peter Maciulaitis, Patrick Shea, Lee Swent, Karen Vendl, Mark Vendl, Bill Wahl, and numerous others.

And how can I begin to thank the hardworking archivists! Michigan Technological University's (MTU) Copper Country Historical Archives has been a tremendous help, thanks to Erik Nordberg and Julie Blair. The Friends of MTU Archives have supported my research twice with valuable travel grants. At the Smithsonian, Shari Stout at the Division of Work and Industry (National Museum of American History, Smithsonian Institution [NMAH]), Craig Orr and staff at the NMAH Archives Center, and the staff of the Smithsonian Institution Archives have shaped this work from the beginning of my research. Paul Coyle at the National Mine Map Repository was terrifically helpful. Jeremiah Mason and Jo Urion at the Keweenaw National Historic Park were fantastic, and helped shape my thoughts on drafting and using maps. The Internet Archive (archive.org) has been instrumental in helping me gain access to major runs of some technical publications; their vision of information-for-all without a profit motive is worth supporting. Thanks also to Ginny Kilander at the American Heritage Center of the University of Wyoming; Rachel Dolbier and D. D. LaPointe at the W. M. Keck Museum of the University of Nevada–Reno; Julie Monroe, Garth Reese, and Marilyn Sandmeyer at the University of Idaho Special Collections; Nelson Knight of the Utah Division of State History; Debbie Miller at the Minnesota Historical Society; Shawn Hall, former director of the Tonopah Historic Mining Park; Eva La Rue of the Central Nevada Historical Society; and the staff of the Columbia University Rare Book and Manuscript Library for going out of their way to assist me.

In addition to the Miller Fellowship and the Travel Grants from MTU Archives, portions of this work have been assisted by the Faculty Development Fund of the College of Liberal Arts at RIT, a predoctoral graduate fellowship from the Smithsonian Institution, and Johns Hopkins University.

Matt Becker, my editor at the University of Nevada Press, has been instrumental. His enthusiasm for the project, and for mining history in general, has made it easier for me to overcome self-doubt, and his wise advice has definitely improved the result. I also owe thanks to the press's anonymous reviewers, who had many helpful suggestions, and the creative and diligent staff who ushered the book into print, especially Mike Campbell, Alison Hope, and Kathleen Szawiola. Hyungsub Choi, Brian Frehner, Tulley Long, Allison Marsh, and Andy Russell have been invaluable sources of ideas, motivation, and support, and I can only hope that one day I can help them as much as they have helped me.

My family has been an essential bastion of support through grad school and writing the book. My parents, via blood and marriage, have always cheered my success: Mary Jean Nystrom, Ed Land, and Ann Marie Land. My sisters have both contributed directly: Gretchen Higgins, a photographer, helped me with several images, and Melissa Salmon and her family put me up for a week of writing and home cooking during a research trip in 2010. More than anything, Rachel Land Nystrom has made this happen. She has put up the funds needed to finish research trips, encouraged me when I needed it, edited the entire thing more than once, and had more discussions about underground maps than any spouse is obligated to endure. In these and a thousand other ways, she's made my life wonderful and this book a reality. I love you, darling.

SEEING UNDERGROUND

Introduction

In the summer of 2002 the world's attention was riveted to a field in south-western Pennsylvania. On July 24, 2002, nine bituminous coal miners were trapped underground in the Quecreek Mine by a flood of water that was released when the workers unexpectedly broke into an abandoned, flooded mine next to it. Rescue efforts began almost immediately in hopes that the miners might still be found alive. The mine's surveyor, on the surface, used his carefully surveyed underground maps to choose the surface point that corresponded to the most likely location of the crew below. The crews drilled a small hole to provide fresh air and to try to create enough air pressure in a high part of the mine to prevent the miners from drowning. The surveyor's educated guess was correct. Once the hole was drilled, the trapped miners below rapped on the bit to indicate all nine were alive. Then the operation moved into rescue mode. Operators attempted to de-water the mine with huge pumps, and used an enormous drill rig to slowly create a hole large enough to lower a special rescue capsule to remove the miners to safety. Seventy-seven hours and a couple of broken drill bits later, the miners were all rescued alive.[1]

The Quecreek rescue miners and rescuers were lauded as true heroes, and there were no obvious villains. The problem was that the adjacent abandoned mine, called the Saxman, actually extended closer to the boundary line than anyone working for Quecreek realized. When the miners broke through, the Quecreek Mine maps indicated that they were still three hundred feet from the abandoned Saxman workings. Abandoned mines dot the American landscape, and pose a real hazard to subsequent mining operations due to risks of flooding, instability, and explosive gas. Since the 1960s federal and state programs have collected and microfilmed maps of mines periodically, but despite regulations requiring submission of maps, many abandoned mines have outdated maps or no maps at all. In order to get a permit to open the Quecreek Mine in 1998, its operating company had to conduct a thorough search for maps of the adjacent Saxman Mine, to try to determine the exact location of its workings next to Quecreek. The company found several maps, but "most of the maps were antiquated and of no practical usefulness."[2] Two maps, however, could be used. One had been stored at the federal mine map repository in Greentree, Pennsylvania; this map, dating from 1957, was also on file at two state mine map depositories. It had not been marked "final," however, by a certified engineer. By contacting the former owners of the Saxman Mine, Quecreek officials had been able to find a map that dated to approximately 1961. This one was also not certified by an engineer, but it showed workings that were not on the 1957 map. This 1961 map of the Saxman was used by Quecreek to develop the map and plan of mining that composed Quecreek's state mining permit application. The state, for its part, double-checked to make sure that Quecreek had looked for maps in all of the conceivable places, found nothing unusual, and approved the permit in 1999. Mining began in 2001.[3]

Ironically, however, another, better map of the Saxman Mine *did* exist. In June 2002, a month before the Quecreek accident, the granddaughter of the last state mine inspector of the Saxman Mine donated her grandfather's personal papers to the Windber Coal Museum in Windber, Pennsylvania, some twenty-five miles northeast of Quecreek. Included among his effects was a 1964 map of the Saxman, marked "final," though it had not been certified by an engineer. Ominously, the map, shown in figure I.1, depicted in yellow a set of parallel gangways proceeding directly toward the Quecreek boundary, well into space that had been shown as unmined in the earlier maps. When investigators of the Quecreek accident discovered it in August 2002, the map was "found lying in a corner of the museum's attic and was not catalogued or indexed." The map had indeed been provided to the Pennsylvania mine inspector, but had "inexplicably" not been entered into any of the federal or state map repositories.[4]

Figure I.1. The 1964 map of the Saxman Mine, found in a local historical society, depicted tunnels not shown on earlier maps. These excavations, jutting down from the main workings in center-left, filled with water and flooded the neighboring Quecreek Mine when miners inadvertently broke through. Document 39230403, Office of Surface Mining, Reclamation and Enforcement's National Mine Map Repository, Pittsburgh, PA.

The Quecreek disaster and the subsequent discovery of a better mine map in the local historical museum highlight the central importance of mine maps to current mining practice. But how did they become so important? How did they change over time? The search for answers led to mine maps as well as other materials. As industrial mining came of age in the United States in the late nineteenth and early twentieth centuries, mining engineers created a visual culture of mining—a set of practices, artifacts, and discourses tied to visualizing underground mines. This visual culture helped render underground mines—as spaces and as businesses—more predictable, controllable, and understandable. It contributed to changes in the way mining work was done, and who had authority to do it. It also set mining engineers apart, as a professional group, from others who toiled at mining.

The question of visual culture has been fruitful for scholars in many fields since the so-called visual turn of the early 1980s. Generally speaking, these studies used visual evidence alongside more-traditional textual historical sources, gaining insight from visual texts that often cannot be uncovered any other way. At the same time, scholars critiqued the production of the visual sources themselves. Photographs and maps that were perhaps once read only for their visual content were now beginning to be understood as artifacts or productions of culture, with the presumptions of their producers embedded in them, directly shaping the information they displayed.[5]

Underground maps and models, fruits of the visual culture of mining, can certainly be analyzed in this first fashion. For instance, a casual look at a mine map reveals its content, showing what a particular mine—likely now inaccessible—looked like and where it was located. Historic mine maps can still offer useful information of this sort, helping ensure, for example, new subdivisions are not constructed over invisible tunnels, or, more vividly, ensuring that Quecreek-style breakthroughs into abandoned workings are avoided. A trained eye might see pictures of different styles of reaching ore and working the mines, as older methods gave way to newer techniques. In all of these cases, the denotative visual content of the map or model is the focus.[6]

There's more we can understand from these maps and models, however, if we adjust our focus and consider them not only as carriers of visual content, but also as material artifacts. In this context, visual culture is an analogue to material culture—that is, the physical artifacts that were the products and producers of a visual orientation to work and life. Thinking of visual culture in this way highlights the importance of these maps and models as *objects,* and helps us ask questions about their life as objects—creation, use, movement, cost, destruction—that are quite independent from the information they portray.

A particularly important component of the visual culture of mining were maps and models of underground mines—as the Quecreek disaster reminds us. These technical representations played an important role in the formation, maintenance, and power of this visual culture of mining, in two different ways. On one hand, maps and models, produced by engineers, were products of the visual culture of mining. At the same time, however, maps and models were also important components of that culture, key to creating and perpetuating it. The form and content of maps and models varied between sectors of the industry and changed over time, reflecting gradual adoption, uneven professional training, and shifting norms within the profession. The intent of this book is to explore the development of the elements of this visual culture and its impact on the American mining industry in the late nineteenth and early twentieth centuries.

First, a word about the broad arc of mining history in America and how it relates to the development of the profession of mining engineering: People across the globe have engaged in mining as an economic activity since ancient times. In North America the desire to dig mineral wealth from the ground arrived with the first European settlers from Spain and England—Coronado was searching for the famed cities of gold on his trek through the Southwest, and the settlers of Jamestown, Virginia, expended fruitless effort on finding gold and other metals. Colonists mined copper at East Granby, Connecticut, for more than six decades before the Revolutionary War, and colonial miners also extracted copper in New Jersey. Colonists pursued mining of iron (sometimes from bogs) with varying levels of success, from initial efforts in the seventeenth century through the colonial period, until, as T. A. Rickard noted, "by the time George Washington became President, the making of iron on a small scale was established in every one of the thirteen States of the Union."[7] Lead was mined in the region near the upper Mississippi River from the eighteenth century, while the territory was still under French control. Bituminous coal was mined beginning in the 1700s in Virginia and western Pennsylvania, and the anthracite coal of eastern Pennsylvania was mined as a marketable commodity beginning in the first decades of the 1800s. The native copper deposits of Lake Superior were commercially exploited beginning in the mid-1840s, and the iron mines of Marquette commenced production within a decade. Gold mining, also on a small scale, took place in North Carolina from the end of the eighteenth century, and began in earnest in 1829 with a rush to Georgia. The Mexican inhabitants of California mined gold prior to James Marshall's discovery of gold in a millrace in January 1848, which set off the justly famous California Gold Rush of 1849.[8] That rush—which produced our name for those miners: forty-niners—was of vastly larger scale and scope than any of the mining that had occurred in America before. The number of people who became interested in mining as a result of the California Gold Rush helped accelerate exploration for gold (and silver) in other parts of the American West. Major finds were made in 1859 in western Nevada at the Comstock Lode, and in Colorado near Denver, but thousands of smaller mining districts were discovered in the several decades following the Gold Rush.[9]

Much of this early mining, including the California Gold Rush, sidestepped the sorts of problems mining engineers would tackle in a later period. Miners sought only the richest ore, located at or very close to the surface of the earth, and avoided ores that posed complex metallurgical questions (such as certain types of combinations with other elements). As long as mines were not dug too far into the earth, miners avoided the need for pumping out mines below the water table. The gold mines of the South and of California were mostly *placer* mines, where the gold was mixed with gravel and dirt in the ground, rather than trapped inside solid rock. Placer miners used running water and gravity, together with a variety of simple devices for agitating the two, to separate the gold from the mix. In other cases, miners dug underground, but the closer they remained to the surface, the easier the core problems of mining—access, excavation, pumping, and support—were to handle.

None of this should suggest that mining was dead simple, though many of the forty-niners who headed to California seemed to think it was. Both placer mining and underground mining were akin to a craft, best practiced or at least directed by those with long experience and expertise. This expertise came from many quarters of the globe. Experienced European miners, especially those from England, France, and the German states where underground mining had been practiced for centuries, were a key source of talent in Pennsylvania coal mines and underground works elsewhere. In the West, miners frequently used accumulated Spanish and Mexican mining techniques as a starting point. By the time of the California Gold Rush, miners with experience in earlier mining districts (such as the Georgia gold fields) could also serve as a corridor of transmission of techniques and technologies.

Following the first years of the California Gold Rush, American mining moved unevenly from a proto-industrial activity to an industrial one, in a rough synchronicity with the overall pace of industrialization in the United States. In industrial mining, miners typically worked not for themselves, but for a company. Industrial mines were much more technologically intensive and usually more productive, using machines and tools to help miners excavate more deeply and systematically, and to handle greater varieties of ore at a profit, than had been the case in the past. The designs of many of these machines, and the expertise needed to use them, were imported from Europe, but the unusual challenges and opportunities posed by American mines soon prompted the development of a wide array of experimental variations on the classic mining tools, resulting in some inventive triumphs as well as some complete wastes of money and effort.

The scale and scope of the American mining industry accelerated along with industrial growth beginning in the 1860s. Mining companies made mines deeper and more expansive, discovered and exploited more mining districts, and sunk ever-larger capital investments into the earth. Precious metals—gold and silver—represented capital and wealth pulled from the ground. Copper, iron, and coal mines provided raw materials for indus- trialization at a prodigious rate, as American mills and factories converted them into usable goods. By the 1920s, when this study ends, the American mining industry was largely mature and had exported its technologies and expertise around the world.

Underground mines varied substantially in their shape and scope over time, dependent on factors including the mineral being mined, the regional traditions of the mining district, and the relative complexity of technology available at a particular time, but all such subterranean produc- tion facilities shared a handful of common characteristics as workplaces. Of primary importance was that miners were dependent on technology to make habitable the place that they worked. Rosalind Williams noted the power of the underground as a metaphor for an entirely artificial envi- ronment, wholly dependent on technology.[10] Such a picture was not far from the truth. Ventilating technologies served both to bring fresh air into the mine and to remove explosive gases. Roof support technologies such as systems of timbering and pillars of native rock were utilized to keep the mine from collapsing. After mines went below the level of the local water table, pumping technologies were required to drain water from the workplace. Transportation technologies were necessary to move vertically to and from the surface, and horizontally within the mine. One distinc- tive characteristic of underground mines was their darkness; miners were dependent on hand-held or cap-mounted candles or oil lamps to provide what little light was available. Portable electric lights were first used in the United States in 1908, but miners only gradually adopted them.[11] All these individual technologies functioned as parts of a technological system —that is to say, the mine itself.[12] Literally embedded in the earth, an industrial mine was a spatially organized technological system, designed by the mining engineer, excavated by miners, created for the purpose of the extraction of raw material to be processed on the surface into economically useful matter.

The overriding concern of mining engineers was to increase the predictability and control of this system, with an eye toward creating profits for the owners. This was a never-ending effort. Tunnels and holes deep beneath the surface of the earth are very hazardous places for humans to be. The earth might move at any time, the air might lack oxygen or fill with explosive gases, big heavy things could crush fragile bodies, and it is dark. *Really* dark.

Mining engineering as a profession developed in America as mining industrialized, and indeed the engineer was a "focal figure" in this transformation. Historian Clark Spence noted that American mining engineering began that time period with a largely empirical approach, augmented by a handful of European-trained engineers. This early class of mining experts was epitomized by the mining engineers of the Comstock Lode during its heyday in the 1860s and 1870s. Gradually, a larger proportion of American mining engineers were university trained, most of them at the American schools that opened in the last decades of the nineteenth century. The burgeoning opportunities created by the booming American mining industry encouraged graduates of the new American programs to have a wide knowledge base, with a mix of theoretical and practical training. The new challenges posed by mining operations in the American West, in particular, encouraged these engineers to use their scientific and practical training in equal measure to solve mining problems. These mining engineers were jacks of all trades, flexible, innovative, and prepared for and capable of anything.[13]

The impact of this new class on the developing mining industry was transformative. Historians noted that trained engineers redesigned the technological systems at the heart of mining enterprises, resulting in gains in business stability and industrial productivity. The miner, who had been the most important figure in mine work, was displaced by the engineer as the locus of decision-making as the new organizational and technological systems promulgated by mining engineers fragmented the traditionally wide range of skilled work performed by miners into a series of semiskilled and unskilled occupational niches. Pressure created by the depletion of the richest deposits, increased responsiveness to market forces, and, most of all, "the progressive and widespread adoption" of new mining and milling techniques (especially the shift toward less-selective "mass production" mining) powered these transformations that that featured mining engineers as the beneficiaries and agents of change.[14]

How did mining engineers gain power and authority to exercise control over mining operations? It was not simply a function of greater expertise, especially in the proto-industrial period, as the most expert miners were clearly those who had worked longest and experienced the widest variety of situations—not those who learned about mining in the safety of a classroom. Instead, the professionalization of mining engineers, and their resulting power, can be understood as a broad competition, against experienced miners, foremen, and bosses, for authority over the work that took place underground.[15] University-trained mining engineers were at a disadvantage initially because underground spaces were so complex that it took years of exposure to learn even the basics of how they worked. However, mining engineers, as a professional group, wielded advantages in this contest, the most powerful of which were the key elements of the visual culture of mining—visual representations.

Engineers of all sorts have long made important use of visual knowledge and visual thinking. Whether simply reasoning or imagining in three dimensions, or using specialized drawings to communicate those visions to other engineers, these professionals employed their "mind's eye" in a way that is central to engineering work.[16] Engineering drawings put much of this visual thought into paper forms, which then serve as communication devices, both between engineers and from engineers to non-technologists. Terming them "technological representations," historian Steven Lubar described the "enormous power" embodied in these drawings, which can carry and communicate tremendous amounts of densely packed information, can stand in as facsimiles of technological artifacts too large to manipulate at full scale, can serve as focal points for organizing technological systems and technical communities, and can portray scientific truth with a visual rhetoric of substantial persuasiveness. In other words, visual representations are powerful tools in a social context as well as in a technical one.[17]

Mining engineers frequently made traditional engineering drawings in the course of their work, but the visual representations at the heart of their intellectual enterprise were maps and models of the underground spaces that they created. Mining engineers worked under a particular professional handicap—they could not directly see the systems and spaces that they were making. Unlike a nineteenth-century mechanical or civil engineer, who could see the fruits of his engineering vision as a casting or a bridge, mining engineers had no way to grasp the entirety of a mine except through maps and models. As a result, mining engineers had to rely heavily on these visual technologies to understand and control subterranean landscapes—controlling the mine on paper, at least.[18]

The power of technological representations gave power to the visual culture of mining that was formed out of them. These technological representations were the constituent elements of a distinctive visual culture of mining engineering, one that American mining engineers used to try to solve problems and increase predictability in the technological systems they created and supervised. The individual elements of this visual culture—actual maps and models—served mining engineers as tools, potentially useful for multiple tasks. They were made and used in accordance with an evolving set of engineering concepts and practices, which could and did vary widely across American mining engineering. The visual vocabulary employed on these maps and models changed dramatically over time, influenced by changing technologies, evolving uses, legal regulations, shifting educational standards, traditional practices that varied by region, and pressure resulting from the growth of a coherent professional identity.

The development of this visual culture of mining was intertwined with the gradual shift of the power of decision-making in mining from those who worked below ground to university-trained mining engineers, and managers in an office. Looking back on this shift, engineers prevailed largely because they could deliver greater amounts of finished product at lower cost. This was expressed in many ways—being able to plan for the life of a mine more effectively so capital could be acquired and spent to sustain operations, reorganizing work practices to increase efficiency and reduce cost, and using science to aid the search for new deposits and the efficient conversion of raw ore into salable metal. In each of these endeavors, the visual culture of mining shaped the actions of mining engineers—and thus the outcomes.

To explore this visual culture of mining, this book is divided into two parts about underground maps and mine models, respectively. Part I, Mine Maps, begins with an overview of the history of underground mapping to demonstrate how these maps were surveyed, drafted, used, and stored, and the implications underground mapping had for the emerging professional identity of mining engineers. Next, we narrow our focus to the Pennsylvania anthracite coal fields in the late nineteenth century for a closer look at mine mapping there. The anthracite region was a hotbed of mapping innovation because of a complex mix of factors that included the influences of a pioneering mine safety law as well as access to markets, capital, and engineering talent that collectively served to advance mapping practice rapidly in the nineteenth century. The maps themselves were rapidly evolved as tools, which we will see by looking at specific examples from one mine. Part I concludes by turning attention west to Butte, Montana, where experts working for the Anaconda Copper Mining Company developed a system of geological mapping around the turn of the twentieth century that revolutionized mine mapping practice. Beginning by reimagining the form and content of underground maps, the Anaconda mining engineers and geologists developed a new set of visual practices and artifacts that gave them a better grip on control over the complex Butte underground. At the same time, the new emphasis on geological knowledge promulgated at Butte served to adjust the jurisdictional boundaries over underground work yet again, with traditional mining engineers losing ground to mining geologists.

Part II of the book, Mine Models, moves from underground maps to three-dimensional models. Models were visually powerful in ways that ordinary mine maps were not, but were also more expensive and difficult to make and update, which meant that these elements of the visual culture of mining were most useful in places where mine maps fell short— in places where communicating to a nontechnical public was of paramount importance. We take a broad overview of the history and different forms of underground models, and describe each of the genres of mine models; each genre carried different strengths and weaknesses, trade-offs between the persuasiveness of information, technical accuracy, difficulty of construction, and size or weight. Some were used in educating future mining engineers, but many were made for lawsuits, where communicating mining and geological details to a nontechnical judge or jury was worth the trouble and expense of model-making. We will examine one such case in detail: The Jim Butler Mine and the West End Consolidated, two neighboring silver mines in Tonopah, Nevada, fought a high-stakes lawsuit in 1914. Both sides marshaled the visual culture of mining to help them control the legal narrative; a close look at the proceedings of the trial shows how strongly the visual tools affected the outcome of the case. Part II concludes with a consideration of the role of the visual culture of mining in educational settings. Mine models became recognizably important to educating future mining engineers in universities across America, though the question of the best means of obtaining and using them was widely debated. Models also spoke directly to the public in popular exhibitions and museums. A close look at the case of mine models in the United States National Museum, a unit of the Smithsonian Institution, shows how company-made displays made their way from commercial exhibits to the museum halls at the direct invitation of curators and museum brass in the 1910s. Museum officials hoped to promulgate the values of mining engineers to a public audience that needed, in their view, to be more appreciative of the work of the industry.

The overall aim of this book is to convey an understanding of the visual culture of mining in order to shed light on a key element of the history of industrial mining in the United States. It is worthwhile to study the history of the mining industry in detail because of the impact that mining has on everyday life. Mining advocates love to point out, often in the form of a bumper sticker, "If it isn't grown, it must be mined," which of course is true. The consequences of mining, however, can't be distilled into a pithy quote. The enormous natural resources of the United States contributed materially to its emergence as a world power in the late nineteenth and early twentieth centuries. A domestic iron and steel industry used coke made from bituminous coal and anthracite coal, together with vast deposits of iron ore, to facilitate the building of railroads, ships, bridges, and machine tools. Domestic manufacturing also led to an increase in consumer goods such as sewing machines, bicycles, and automobiles. An increased monetary supply, in the form of gold and silver, helped expand capital markets and spur economic growth, but also contributed to inflation and speculative panics, especially those of 1873, 1893, and 1907. Copper from mines in Michigan and the American West formed the wires that connected the country with telegraph, telephone, and electrical lines.

The human and environmental cost of all of this mining was enormous. Coal mining had one of the highest fatality rates of any job in America during the period we will examine, from 1870 to 1920, and metal mining was also quite dangerous. Even in the modern era, the story of the Quecreek miners' entrapment by a flood of water from the abandoned mine next to them was unique not because of the event, but because of their survival. Statistics did not take into account chronic diseases, such as silicosis and black lung, that shortened lives and agonized sufferers. Environmental consequences could be huge. Water that drained from mines was frequently acidic, killing fish and reducing plant life along the banks of the waterways into which it flowed. Sometimes dirty water from coal operations would be impounded behind a dam, in order to allow the particulates to settle, creating the hazards of catastrophic flooding downstream if the dam was breached. Mining operations tended to clear hillsides of timber and stabilizing growth in order to get at the rock and to have a place to put it, but this exposed hillsides to massive erosion, removing topsoil on the hills and silting up streams and rivers in the valleys. Sometimes this was a deliberate strategy, as with the hydraulicking method of mining in California, which used water from high-pressure nozzles to erode entire hillsides. Due to the incredible damage caused by the flow of sediment downstream, this practice was banned in 1884, but other similarly damaging practices continued unabated. Emissions from lead and copper smelters destroyed trees and adversely affected human health in the Rocky Mountains and elsewhere. Processing mills, especially early mill designs such as those used on the Comstock Lode, allowed reagents including mercury to wash into rivers and gulches. And large-scale open pit mining, in common use for just over a century, devoured communities and left enormous toxic scars on the landscape. To the extent that mining engineers have been the primary architects and builders of this mined world, it is important to understand their work in order to have a better appreciation for the world they have left for us—for better and for worse.

PART I

Mine Maps

chapter 1

Underground Mine Maps

Maps are powerful because they allow us to learn or know about places that might be far from our present experience, but in a way that relates the unknown to other places on the earth. They organize and represent space visually. However, most people don't need a map of their hometown in order to find their way around. Through long exposure, we usually build a map in our heads that allows us to navigate more efficiently; physical maps are, in that sense, not a prerequisite to finding our way.

The same was true with underground mines in the preindustrial era— maps were rare and largely useless, of little importance to mining operations. The veins led the miners, and the works evolved slowly enough, in the era before high explosives and mechanical drilling, that keeping track of underground progress was not difficult. Mines of that era rarely had problems that underground maps could help solve. However, as mining became industrialized in the late nineteenth and early twentieth centuries, maps moved from nonexistent, to an afterthought, to an important way of portraying the mine, to a key managerial tool. Underground maps were the most central element of the visual culture of mining, created and wielded by mining engineers in the era of industrial mining in the late nineteenth and early twentieth centuries. This shift is rooted in the visual properties of maps. Maps are powerfulthings.[1] Mine maps were not like those that historians of cartography have ordinarily studied, however. Because they were unpublished, made by the application of mechanical drawing techniques, and created by engineers for engineers, underground maps share some of the same properties as engineering drawings—the "technological representations" that help engineers wield power and control, as explained in the introduction.

Chapter 1 will provide a broad overview of underground maps, looking in detail at how they were made and used, and showing some of the changing forms of mine maps over time. The chapter will also explore how mapping became an important element in the education of mining engineers, and how it grew to be a central part of the professional identity of mining engineers as they competed with other groups for jurisdiction over underground work.

The first step in creating or revising an underground mine map was to perform a survey. Surveying was the process of using instruments to measure angles and distances; engineers could later use this information to measure and plot a map of the mine at a reduced scale on paper, using geometrical principles.

Maps of surface landscapes have been associated with mining for many centuries, including some of the earliest maps associated with mining in the United States. Such maps, made with the same tools and techniques as other surface maps, show the surface location of mineral deposits or land ownership claims. Many Western states required a surface map of a claim to be made by a certified US deputy mineral surveyor before the claimants could transfer it to private ownership by patenting the claim. While such surface maps were important, they did not require specialized personnel to create or use them. If an ordinary land surveyor could make a map that included a house or a stand of timber, the same surveyor could note a mining property line or mineral outcrop. Certification of deputy mineral surveyors was important not necessarily because they possessed extraordinary skills, but to help reduce conflict over claims surveyed by competing companies. As a result of the relative simplicity of these surface maps, particularly in the early years, they contributed little to the development of the visual culture of mining.

In addition to surface property and claim maps that depicted the surface of the earth, another precursor of underground maps were geological maps that purported to show geology below the surface. The geological sciences coalesced rapidly in the first several decades of the nineteenth century, and the visual representation of geological features played an important role in geological interpretations.[2] William Smith is usually credited with having produced the first geological map, in 1815.[3] Even so, J. P. Lesley, a well-known American geologist, lamented in 1856 that geological maps "are commonly mere distortions and caricatures which illustrate the subject by confusing the observer. Neither maps nor sections are helps to the geologist—and are cruelties to the schoolboy—unless they be properly and accurately made. Then they become his efficient tools. But he must make his own tools. He will find none made to his hand. The age of maps has just set in."[4]

Lesley's 1856 *Manual of Coal and Its Topography* did not discuss underground mapping at all. One significant reason is that there was little mining activity in America in 1856 that was deep enough to map, and few engineers who might have done the work. This situation changed dramatically in the decades to follow, and it started with surveying underground mines.

Underground surveying shared much in common with surface topographical surveying. Both measured areas by noting distance and angle measurements between a series of points designated by the surveyor, as well as known topographical features such as large trees, church steeples, roads, or surveying monuments. Surveyors then plotted these points and measurements on a map.

However, mine surveying introduced several complications into normal surveying practice. On the surface, a surveyor could orient work to true north by observing the sun, the stars, or a natural reference point in a known location such as a mountaintop or church steeple. Underground, none of these options was feasible, which meant that a known reference and orientation had to be carried underground from a surface survey. Underground mines were dark, which meant that instruments and targets had to be illuminated with lamps. The cramped spaces underground often required a dexterous transit operator, and in some cases required alternatives to ordinary tripods to save space. Mines were also hazardous, with uneven floors, dirt, damp, and rocks, all of which made the surveyor's job less pleasant and prone to additional inaccuracy. Finally, the linear nature of underground mines made errors more likely. In surface surveying, an important principle is to create a closed polygon, surveying around the land until the survey ends where it begins. This method gives the survey operator an effective check against mistakes in measuring angles, since the angles of a polygon must add up to a particular value that is dependent on the number of angles, not their degree measurements. If the measured angles are not close to the predicted total, a measurement was off and the survey should be double-checked.[5] However, the shapes of underground spaces made it difficult for surveyors to measure them with a closed polygon, often denying underground surveyors an additional check on the accuracy of their work.

Executing underground surveys required a collection of tools, each with a particular application in the series of steps involved in a survey. These tools, in turn, each had a history of its own; they were gradually refined by their makers with an eye toward cost, accuracy, and speed. Thousands of engineering hands, spread across the country, formed the visual culture of mining with these tools, just as a carpenter's hammer produces a house. Let us examine each of them briefly.

In order to determine side-to-side and up-and-down angles, preindustrial miners generally used two tools. One was the compass, which could measure side-to-side angles in the same fashion underground as above ground. The other was a clinometer, used for measuring up-and-down angles, where a weighted tongue or plumb bob would point toward the ground and the edge of the protractor-like clinometer would represent the angle to be measured. Miners usually used these tools by tightly stringing a stout cord between the points to be measured. They then could place the compass and clinometer along the cord and take the measurements. Later evolutions of compass and clinometer were given gimbals and little arms to hang on the stretched cord, resulting in the hanging compass. As may be imagined, these tools were cumbersome to set up and use, and were not very accurate by later standards; given the limited role of visual information in preindustrial mining, though, the deficiencies may not have been an issue.[6]

The compass, in particular, had inescapable defects and unique advantages. As was recognized above ground, the variation of magnetic north from true north on a regular and unpredictable basis meant that readings in the same spots on different days might yield different results. Advocates of the compass pointed out that, for surveys that could be completed in a single day, the compass was accurate enough, as all readings on a particular day would have the same declination from true north, making them at least accurate with respect to each other. As may be imagined, as industrial mining created mines of multiplying underground complexity, this defect of compasses became more problematic. The other issue with compasses was that they were deflected by the presence of iron nearby. This meant they could not be used in iron mines, for example, and over time, as iron machinery became more common in industrializing mines, the presence of pipes, tracks, and other items rendered them so unreliable they became nearly useless for accurate work.

Even so, the compass had several advantages that kept it in use in more conservative mining regions, such as Cornwall, for many decades after optical surveying instruments had become standard. One advantage was that the compass was fairly simple to understand, and required no more expertise to use below ground than above. A second, important feature was that a compass could provide a bearing below ground to start a survey that was consistent with one taken on the surface (presuming the needle was not deflected by the presence of underground machinery). This allowed the engineer to orient the geometric polygon he created with proper respect to surface features. This feature kept compasses as part of the equipment of many mining transits well into the era of industrial mining, though compass readings eventually served primarily as an additional check against clerical errors.

A major change in surveying equipment occurred in the early nineteenth century, with the development of optical, rather than magnetic equipment. These new instruments were more accurate than magnetic compasses, but were also considered more delicate, slow to use, and—importantly—expensive. Such optical instruments were commonly called theodolites, though many individual inventors had pet names for their own creations; as we shall see in a moment, most mine surveying in the industrial era was conducted with an improved successor to the theodolite, the transit.[7]

Theodolites and transits measured angles on both vertical and horizontal axes. In principle, a horizontal tube or telescope was mounted with extreme precision at its horizontal and vertical pivot, meaning that as the tube was moved from side to side and up and down to bring the target into view, it accurately measured the horizontal and vertical angles on scales attached directly to the instrument. Since there was no reliance on the attraction of a magnetized needle to the earth's magnetic field, such an instrument worked perfectly even where iron was abundant. Furthermore, as instrument manufacturing became more precise, optical instruments, in turn, could provide readings that were more precise, where reliance on the variation of a compass needle would inevitably leave a larger margin of error even in compasses with the finest craftsmanship.

While the theodolite was used widely in railroad and land surveying, and in some mines, especially abroad, the visual culture of American industrial mining rested primarily on surveys made with the transit. Theodolites of that time used a quarter or half circle, measured in degrees, to record vertical angles; the transit by contrast used a full circle, graduated all the way around, and mounted a shorter telescope higher above the base of the instrument. As a result, unlike the theodolite, the transit's telescope could flip completely over in its mountings. This allowed mining engineers to turn their transits over to obtain a set of check readings, ensuring both that the instrument had not been knocked out of alignment and that no mistakes had been made in obtaining or recording the readings.

Several variations on transits were used by mining engineers to solve surveying problems peculiar to mine surveying. Many transits through the end of the nineteenth century were made with large integrated compasses, which could be used to orient the instrument or as a further source of data that could provide a double check of the optical readings. One issue was that mining engineers were much more likely to have to measure steep vertical angles, in measuring shafts or raises, for example, than would a surveyor using a theodolite on the surface. If the transit telescope was mounted directly over the center of the instrument, its base interfered with measuring angles beyond about seventy degrees. Mining engineers and instrument makers developed several designs to cope with this problem, including mounting an additional removable telescope, called an auxiliary, on top of or to the side of the main tube, where it would provide a clear view at a vertical angle. A different solution could be had by setting the transit's telescope off from the vertical center of the instrument on inclined stanchions. An auxiliary telescope can be seen in the Heller & Brightly mining transit, made in the mid-1870s, in figure 1.1.

Whatever instruments they chose, mining engineers used surveying to connect points within the mine, and made maps based on these skeletal measurements. Any type of surveying is dependent on establishing carefully located points of reference on the earth, usually called surveyor's stations. Common examples for ordinary land surveying include nails

Figure 1.1. This mining transit, manufactured by Philadelphia instrument maker Heller & Brightly in the early 1870s, has a telescope mounted on stanchions above a compass, with a vertical scale on the side. This transit also features a detachable auxiliary telescope mounted above, and parallel to, the main sighting telescope. This could be used to measure near-vertical angles, when the primary telescope's view would be blocked by the compass base. The firm's mining transit was also smaller and lighter weight than a standard engineers' transit. From Dunbar D. Scott and others, *The Evolution of Mine-Surveying Instruments* (New York: American Institute of Mining Engineers, 1902), 278.

or stakes driven into the ground; readers may be familiar with the brass plaques, set in concrete, that serve to anchor nationwide geodetic surveys. Underground surveying required a modification of this practice. Mine surveyor's stations could not be established with precision on the floor of the mine because work traffic would disturb them. Consequently, the stations in a mine normally consisted of a horseshoe nail with a hole or a hook (known as a spad or spud) driven into a mine timber or a wooden plug inserted in a small hole in the roof. Over time, it became common as well to permanently mark a station's number in the mine's survey system by attaching a metal tag to the spad or painting the number on the adjacent wall.[8]

Over time, these survey stations became part of a permanent information infrastructure in the mine. Early mine surveyors did not always associate a permanent number with a particular survey station, but over time it became common for them to give each an identifier of some kind. One system was to designate each surveyor's station with a unique number, so no two stations would be the same. This system could be used by inspectors and others not intimately familiar with the mine to help orient themselves. Its drawback, however, was that numbers might not be consecutive as work progressed in a particular area, and one surveying crew had to keep a master list of numbers.[9] By contrast, a competing system numbered stations consecutively from the beginning point of the survey, but added a notation to each that reflected the cumulative distance from the beginning point. This system made it easy to calculate distances between stations on the map or, if the stations were visibly labeled, underground. Though the station numbers seemed confusingly similar to each other at first glance, practice suggested the system was not difficult to understand, reported one engineer who surveyed this way, but in the end he concluded, "[B]oth of these systems are used extensively, however, and mine transitmen do not often wish to change from their first love."[10]

The survey stations were aligned with the mining engineer's equipment using small plumb bobs—pointed weights at the end of a string. In surface surveying, the surveyor placed a theodolite over the surveyor's station, then hung a plumb bob from the underside of the instrument on an eyelet at the very center of the device (at the top of the underside of the tripod). The surveyor then moved the theodolite and tripod until the bob pointed precisely at the center of the surveyor's station on the ground. Once the perfect position was achieved, the instrument was made level using leveling screws. In a mine, where the surveyor's stations were located overhead, mining engineers had two options: They could plumb down from the spad to the floor, then set up a temporary marker called a center-pin, which was a block of lead with a steel pin to represent the surveying point. The mining engineer then removed the plumb bob from the ceiling, put the transit into place, and moved it into position plumbing from the bottom of the tripod to the center pin. When the engineer moved the transit to the next station, he picked up the center pin and repeated the process.[11] The second option was to set up the transit directly below the plumb bob. Over time some transits were manufactured with special markings on the top of the instrument to indicate the precise center, to allow mining engineers to use this faster method of setting up their device.[12]

Typical practice in the industrial mining era involved using three survey stations at once in a line—one in front of the surveying crew, one in the center, where the transit was set up, and one behind the crew. While the transit occupied the middle station, at the other two plumb bobs were run from the spads for the targets. The engineer operating the transit would take aim at the target for each measurement. A target could be as simple as a mine lamp left on the ground under the plumb bob, or a light held by a helper behind the plumb bob string. More-elaborate targets with lamps built into them were also constructed, including one that combined a plumb bob and light in the same device, which permitted the helper to set up the lamp and leave it while assisting some other part of the survey operation.[13]

After the survey crew had made angular measurements with the transit, the crew had to measure the distance between the stations. Before the 1880s crews measured with a surveyor's chain, of the same sort used above ground. To use it for measurements, a crew member drove a long pin through one end of the chain immediately under the starting station. The crew then stretched out the chain, usually sixty-six feet in length and comprising a hundred links, straight in the direction of the next station. When at maximum extension, the crew pinned the far end, took up the first pin, stretched the chain toward the target again, and repeated the process until they had measured the entire distance between the stations.[14] In the 1870s mining engineer Eckley B. Coxe invented the first surveyor's tape measure, with five hundred feet of thin steel (which had originally been intended for hoop skirt bands!), marked every ten feet, wound on a reel. This increased both the accuracy and the speed of surveying, and soon spread to all forms of surveying, and not just surveying by mining engineers.[15]

The surveyor recorded all of the data generated underground—data from the surveyor's stations, the angular measurements from the transit, the measurements from the chains, and, usually, some short, qualitative information about the mine as the crew passed through it in a notebook. Given the conditions underground, mine surveyors' notebooks were particularly susceptible to becoming damaged and dirty. Textbooks on mine surveying cautioned surveyors to use a book with sturdy leather or aluminum covers or, better, to use a loose-leaf notebook so that only one day's pages were underground at a time.[16]

After the survey was conducted, the mining engineer transformed the raw data into a map of the underground. This transformation took place away from the mine, in the mine office building's drafting room. While the entire office building was dedicated to the exercise of managerial power, the drawing or drafting room was the most important space for the creation and use of visual representations. While they were commonly scaled to the size of the mining operation and varied somewhat in their characteristics from company to company, these large drafting and rooms were, in the era of industrial mining, often purpose-built spaces designed to accommodate most of the technologies and labor associated with creation and use of a system of maps, drawings, and blueprints that collectively made up the visual culture of mining. These rooms were usually major uses of space in specialized mining office buildings. They combined open spans with lots of windows to let in light with specialized furniture, especially large tables, and at least some document storage such as pigeonholes or map drawers. While the large drawing rooms were probably the most important spaces, nearby smaller spaces, such as blueprint production rooms, photography darkrooms, and storage vaults, also played a role in the visual culture system. The offices of the managers and engineers were usually located nearby, in close proximity to the heart of the visual document system. Many of the examples in this chapter are drawn from the history of two major mining companies in the so-called Copper Country in the Upper Peninsula of Michigan. The Calumet and Hecla Mining Company (C&H) and the Quincy Mining Company were the two of the longest-lived and most successful companies that mined the native copper deposits of the Keweenaw Peninsula. Fortunately for historians, the office buildings of both companies are now part of the Keweenaw National Historical Park, and much of the physical infrastructure is extant, which provides a material complement to the extensive archival resources of the two companies held locally.[17]

The drafting room contained the supplies needed to enable engineers to translate survey notes from the underground onto a two-dimensional plat. These materials included rolls of thin tracing paper, thicker vellum or tracing linen, and heavier stock for more-permanent maps or drawings; pencils, ink pens, inks of various colors, colored pencils, compasses, straightedges and rulers, dividers, protractors, and similar mechanical drawing apparatus. Paper would typically be cut to the size needed for the job at hand, and any future expansions. It was not uncommon for underground maps to outgrow their original sheets, and have additional paper glued on.[18] Odd scraps or misexecuted blueprints were frequently saved and used for scratch paper or smaller projects. For example, surviving maps from the Quincy Mine show how engineers used scratch paper (in this case, the back side of blueprints with printing mistakes) to sketch out the initial plat of a map or the outline of an item in pencil. Once the basics were established, the drawing was copied in a more finished way to a new sheet of tracing linen.[19]

In the era of industrial mining, large, flat surfaces distinguished rooms where visual representations were made and used. Users of large maps and drawings needed to be able to unroll or otherwise deploy the large sheets in order to read, copy, or create them. Drafting tables, with slightly inclined surfaces, served as oversized desks for company drafters, and large flat tables accommodated larger items. Particularly in the nineteenth century, maps were frequently stored rolled, and drafting desks had pigeon-hole storage in the body of the unit to keep rolled drawings close to hand. As mining operations gradually moved to a greater proportion of flat drawings and maps, storage could be integrated into desks in the form of drawers. The C&H made custom office furniture in their shops that combined a rack of map or drawing drawers with a solid slab top. These units allowed engineers to have a large number of drawings close at hand, stored flat, while giving them large flat spaces that were particularly suited to working with large maps and drawings.[20] One C&H-built marble-topped hybrid table–map case had drawers on both sides and was clearly intended to occupy the center of a room, permitting access to all sides; it reinforced the point that the maps were, literally, central to engineering practice, in the office and at the mines.[21]

Drafting underground maps from mine survey data had much in common with plotting standard land surveys, including utilizing ordinary drafting tools. Drawing commenced in pencil. First, the map maker carefully plotted and connected the survey stations. Using notes and measurements preserved in the surveyor's notes, they could fill in the nearby mine topography on the emerging map to suggest the roughness of mine surfaces, though some map makers preferred the clean lines of a ruled pen.[22] Many maps depicted multiple levels of the mine as though they were superimposed on each other, usually showing each level in a different color to distinguish it from the others. Such a "composite map" was the

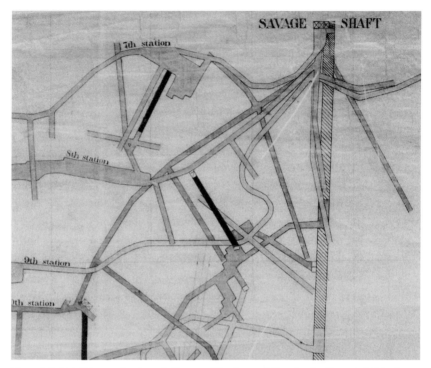

Figure 1.2. Detail from a "composite" map of the Savage Mine, in the Comstock District of Nevada, from the late 1870s. The map depicts the horizontal workings in the mine spreading in all directions from the vertical mine shaft, as though looking down on them from above. Different levels are represented all on the same map, usually distinguished by color. Document 50206126, Office of Surface Mining, Reclamation and Enforcement's National Mine Map Repository, Pittsburgh, PA.

standard way of depicting hard-rock mines in the West, but even anthracite coal maps sometimes superimposed multiple levels of workings. Figure 1.2 shows a portion of a composite map made of the Savage Mine on the Comstock Lode in the late 1870s.

To the basic drawing of the levels, specific features might be added, such as shafts, doors, elevations, or notes about the condition of the mine. Symbols were usually consistent in each company, but many variations across maps from different companies persisted long into the twentieth century. Drafters rarely explained symbols on a map in a legend; the only common exception was when parts of the map were colored to represent a specific phenomenon, in which case drafters would note the meanings of the colors. Drafters sometimes added a title panel, with the name of the map, date of revision, and perhaps a notation of scale. Map scales might be set by law or tradition in particular sectors of the mining industry. For example, surveys in Pennsylvania anthracite mines almost invariably used the scale of one inch to one hundred feet, which the mine safety law governing the mines required. The most common scale on maps of hard rock metal mines in districts across the American West was one inch to forty feet, termed forty scale, but this seems to have been a custom rather than a legal requirement. Many mine maps in the Comstock Lode were constructed with a scale that was a multiple of thirty feet, with either thirty or sixty feet to the inch (as was figure 1.2), likely hearkening to the use of the fathom in measuring mining distances.[23] Whichever scale was used, when the map was complete, it would usually be checked carefully by the mining engineer, if he had not done the drafting.

Early mine maps were normally on heavy paper, often backed by cloth, which was durable and easy to draw on. These early maps were frequently impressive artistic productions, with shaded effects and small flourishes such as tiny profiles of shaft houses. If a copy of one of these maps was needed, translucent tracing linen, similar to modern vellum, was used: the drafter would place the tracing cloth atop the paper map and trace the features that appeared through the semitransparent material.

The prevalence of heavy paper maps decreased beginning in the 1890s, as a new technology arrived in the drafting room—the blueprint machine. Though the change was gradual, blueprinting eventually had a dramatic impact on drafting practice and facilitated the reorganization of the flow of information within a mining firm. Blueprints offered technical and organizational challenges and opportunities to mining engineers and managers, because they were a relatively cheap and fast method of making an exact duplicate of maps, drawings, or orders. The proliferation of blueprints tended to centralize control in the hands of the engineers and accelerated the importance of visual information within mining organizations, because it became significantly easier for documents, and the visual information they carried, to make the transition from the safe confines of the office into the hazardous spaces of the mine below ground, while the original visual representation remained safely in the office vault. This in turn made it far easier to precisely convey the engineers' vision of work to be performed to those who would do the actual labor, and allowed the engineers themselves to take their visual representations with them to help them with decision-making on-site.[24]

To make a blueprint, sensitized paper was placed along with a drawing in an apparatus designed to hold them closely together, and the combination was exposed to the sun or bright light (figure 1.3). Ultraviolet light passed through the drawing and exposed the blueprint paper, but where light was blocked by ink on the drawing, the blueprint remained unexposed. After the exposure was complete, the print was washed in water to remove unexposed material and fix the print, yielding a negative copy of the drawing—white (unexposed) lines on a blue (exposed) background.[25] Since the heavy paper maps were too opaque to duplicate, the arrival of blueprinting also signaled a shift to using tracing cloth as the material for the main mine map. Engineers took other steps to enhance their ability to create blueprints as well, including drawing heavier lines without the artistic shading of earlier practice, and using cross-hatching instead of color to represent information on mine maps.[26] Given the need to expose the entire blueprint at once, the new technology also placed practical limits on the size of visual representations. This provided an incentive for companies to draw their maps and drawings on sheets of a standard size that could be duplicated easily, but that also happened to fit neatly into flat file drawers of standard sizes.

A series of blueprints could replace a copy of a map that had to be updated, and it seems that this was one of the early uses of blueprinting in mining. The early laws made provisions for the mine owner to comply with the update requirements by sending a small sketch of the new work only, or by submitting a verbal narrative of the work that had been done. Later safety laws specified that the inspector's map should be returned to the company for annual updates, then returned to the inspector's office. Several laws, beginning in the mid-1880s, also permitted the submission of "sun print" copies of the map instead of the process of returning the map, updating it, and sending it back.[27] Certainly the production of a blueprint was far easier for the mine owner than was tracing the new work on to an old map or compiling a written account. For the mine inspector, blueprints eliminated the need to send tracings back and forth to all of the mines under jurisdiction, reduced the amount of time that any given map was unavailable, and, even better, almost certainly would have increased the compliance of mine owners with the law requiring updates.

Figure 1.3. A drawing on semitransparent tracing linen would be placed along with sensitized blueprint paper in this solar-powered blueprint frame, and the assembly would be left in the sun to expose the print. This example, likely circa 1890s and located in situ in the attic of the Quincy Mining Company office building, was pushed out the attic window on rails. Quincy Mine Office building, Keweenaw National Historical Park, Hancock MI. Photo by Eric Nystrom, 2010.

Blueprints helped bring visual information in the form of maps closer to the actual working spaces of mines. A 1923 book of sample mine foreman examination questions explicitly placed blueprints at the mine: "Blueprint copies of the mine map should always be available at the mine for inspection."[28] In this quote, it is unclear whether the author is referring to the mine as an underground space, or the mine as an operation, but blueprints could be useful in each place. In cases where a large company owned a number of mines spread out over large distances, such as in the bituminous coal fields after the turn of the twentieth century, the mine maps might be kept by the central engineering office at some distance from each mine. If a local emergency were to occur, such as a fire or a situation where a rescue was needed, the local mine foreman might not have geographic information at his disposal. However, the ease of reproduction of blueprints meant that the central engineering office could produce a blueprint of the main map for each mine, which could be kept at the local site in case it was needed.

Similarly, even if the mine office was located near the shaft, as was the case at the Quincy and C&H, it would be possible for supervisors underground to use blueprinted copies of mine maps to direct work. A supervisor with an underground office could store a blueprint there and consult it frequently, perhaps in concert with individual miners. Before blueprints, if an underground map were to inform the daily business of digging out the mine, the foreman would have had to travel to the mining office, look at the map, perhaps take some notes, and then return underground to deliver verbal directions to the miners. With blueprints available, the foreman could consult with the engineer and receive a blueprint, which he could carry underground with him. The blueprinted map might be able to help him answer any subsequent questions after he left the office, and the foreman could refer to it when giving orders to the miners.

Blueprints might also be useful underground to state mine inspectors and engineers from other companies, who would all be familiar with the visual language of mine maps. Mining engineer W. L. Owens described the use of a blueprint underground for orienting a visiting engineer. "If there is someone making an inspection of the mine who is not familiar with the different headings, he can immediately locate himself by noticing a station number painted on the rib and picking out that number on his blueprint of the mine."[29]

Whether underground or merely at the mine site, the blueprints would ensure that those performing (or supervising) the mining work hewed more closely to the engineer's plan of what was to be done, because the blueprints could communicate the visual component of that plan to the site.

Blueprinting eventually wrought significant changes in the engineering labor force at mining companies as well. The work force in drafting offices before the advent of blueprinting tended to be young engineers at the start of their careers. They would begin by doing relatively menial tracing work, and move up to positions of greater responsibility, even as they became familiar with many aspects of the operation through interaction with the company's corpus of visual culture. Blueprints reduced the number of drafters needed because operating the blueprint machines could be done more inexpensively by less-skilled labor, such as boys.[30] A blueprint, if printed correctly, was as truthful as the original drawing, but if it was incorrect, it was obviously wrong and could be done again. On a tracing, errors could seem like truthful data. Thus tracings needed to be executed (or double-checked) by someone capable of understanding the intellectual content of the work, where blueprints only needed to be proofed for proper exposure and readability. The shift to electric blueprint machines from the frames used to expose prints in sunlight reduced the labor force even further, as one boy could do more blueprinting work than two or three had been able to accomplish with the solar frames.[31] From work practice to the constitution of the office labor force, blueprint technology served to spur gradual but profound organizational changes that tended to increase the decision-making power of the managers and engineers at the top, at the expense of some of the autonomy of employees who worked for them.

After a map, drawing, or blueprint had been made, it needed to be stored in order to protect it and facilitate its easy retrieval. Just as American businesses were addressing the need to store correspondence with a variety of filing technologies, visual representations were stored in several different ways. Furthermore, individual companies frequently used more than one means of storage, depending on the particular map or drawing's characteristics.

Pigeonholes were most useful for storing maps and drawings of varying sizes. Pigeonholes were commonly used before the turn of the century for filing letters and other business papers, and were built into many desks.[32] Pigeonholes for maps merely needed to be a bit larger and deeper. Larger pigeonholes were used for company storage systems, especially for extremely large maps. The second story of the fireproof vault of the Quincy Mining Company office building contains a set of large, deep pigeonholes for storing maps and drawings. A slightly different way to store rolled maps, utilized by both Quincy and the C&H, involved putting rolled maps in metal or cardboard tubes and placing them horizontally on special racks attached to the wall. The tubes or maps were then labeled on the outside. Some of these tubes might have been the light-proof metal tubes in which commercially prepared blueprint paper was sold. The tube storage system protected the maps as well or better than ordinary pigeonholes, and was more space efficient, since it did not require a deep space.

Flat drawers, holding flat sheets, could store more maps and drawings in a given space than was possible with systems such as rolled maps stored in pigeonholes or tubes. The big drawback to flat drawers was their inability to hold visual representations that exceeded the dimensions of the drawers. As the advent of widespread blueprinting in the late 1890s began to restrict the maximum size of new visual representations, flat drawer filing solutions for maps and drawings became more useful. A big advantage of flat drawers was that they could be built into other drafting room furniture, as was done at the C&H, thus permitting maps and drawings to be stored close to where the engineers worked. Mining engineer H. L. Botsford described using flat drawers of several sizes, so that each size of drawing, standardized within his mining firm, would be placed in a drawer with others of its size. Botsford filed all "maps, tracings, blueprints, etc." together according to size, but others, such as engineers at the C&H, segregated tracings from blueprints.[33]

A further consideration about storage of mine maps was fire. Most mining office buildings built for the purpose had a vault or vaults for storing payroll as well as important business documents. Especially given the remoteness of some mining operations, the office vault was considered the best means to defend information critical to the company from fire. Maps and drawings were, by the late nineteenth century, certainly included among the ranks of critical documents. Reproducing the content on older maps might be impossible, as the workings would become inaccessible over time due to flooding or cave-ins, leaving mine maps and old survey notes as the only records of these underground spaces.[34] For example, the Quincy mine office vault was two stories, located at the center of the building. The business operations of the company, including the financial operations such as the paymaster window, were handled on the first floor of the building, and the first floor of the vault was set up mostly to protect books, cash, ingots, and similar materials. Engineering activities took place on the second floor of the building, and the second floor of the vault was built to hold maps and drawings, with pigeonholes, racks for rolled maps, and flat drawers built directly into the vault.

Vaults emphasized both the importance of the visual culture of mining to mine management as well as the trend, spurred by blueprinting, of centralization of information in the form of a master copy. If any hazardous work that a drawing needed to do could be done by a destructible blueprint copy, then the master could remain in the office vault, safe from fire and other hazards. The master needed to be removed from its stronghold only to be updated with the latest information, and to reproduce more blueprint copies to carry the information into the field.

Storage systems were largely useless without some form of organization. The simplest organization was having maps kept by the person who needed them—desks with pigeonholes or drawers are classic examples. As long as there were not too many maps or drawings in use, the simplicity of such a system outweighed the cost of time it might have taken to find any particular drawing when it was needed.

As the number of visual representations in use in mining operations increased, due to both the increasing activities of engineers and the ease of duplication afforded by blueprinting, the organizational problem of storing and finding those representations increased dramatically. Many mining companies struggled to find a system of organization that was flexible enough to accommodate multiple copies, multiple storage locations, multiple types of documents, and multiple uses, while not creating a system of organization so complex that it could not be applied rapidly, be used universally, and be understood by its users. A vast number of surviving visual representations from mining companies bear the imprint of having been part of multiple overlapping organizational systems.

Both the C&H and Quincy mines eventually developed sophisticated card catalog systems to control their drawings. At least for the Quincy, the implementation of the card catalog system prompted a company-wide reorganization and renumbering of the company's visual representations. Quincy kept two card catalogs: one tracked the maps and drawings by the sequential number they were given upon being entered into the system, and the other filed cards by topic, possibly reflecting the spatial organization of the drawers in which they were stored. Both Quincy and C&H relied on the card catalog, rather than the number on the map, to provide the metadata (to use a modern term) that situated the visual representation in its work and storage context.

A related system was used by Canadian mining engineer H. L. Botsford, in which every visual representation was given two numbers—a serial number and an index; Botsford used twin card catalogs, one numerically sequential and the other organized by topic, to find the drawings. The serial numbers inscribed on each visual representation were sequential across all drawings, much like those of the C&H or Quincy. The index number, by contrast, described where the drawing was filed: "A-4," in Botsford's example, would be the fourth drawing from the bottom in drawer A of his flat files. As a result, Botsford could file drawings together that were nonsequential in serial numbers, which allowed him to group drawings by size, topic, or location.[35] He would also know, without looking at the index, where to put away a map that had been taken out.

Not all filing systems chosen by mining companies worked in this fashion. By contrast, the last organizational system used by the Coxe Brothers & Company, an anthracite coal mining firm, used a very sophisticated numbering system with three levels of hierarchy in addition to a sequential map number. Thus, if one knew the system, the map in figure 1.4 could be identified from the tag alone. The "D" signified that the map covered the Drifton Colliery; class 1 was "Land and Surface Maps"; subclass 9 embraced maps of "Railroads, Creeks, Canals, Telephone Lines, Transportation Outside"; this was the twelfth map of in that particular class and subclass for Drifton; and the map was stored, rolled up, in drawer or pigeon hole 205. The information was also kept in a card index, but the system was clearly designed so the cards would not be relied on for retrieving the correct map.[36]

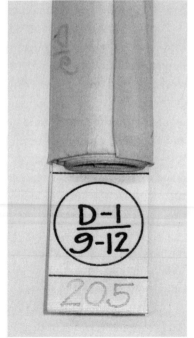

Figure 1.4. Tag on the end of a rolled map from the Coxe Brothers & Company. This tag provided information about where the map should be stored, as well as its contents, to anyone familiar with the coding system. In this case the map was from Drifton colliery (D): class 1 was "Land and Surface Maps"; subclass 9 embraced maps of "Railroads, Creeks, Canals, Telephone Lines, Transportation Outside"; this was the twelfth map in that particular class and subclass for Drifton. End tags allowed rolled maps to be stored in pigeonholes. Division of Work and Industry, National Museum of American History, Smithsonian Institution. Photo by Eric Nystrom, 2005.

Whatever filing system was used, engineers needed to be able to locate the maps and drawings. At C&H, a map's index card gave its drawer number. In that company, drawers were numbered sequentially, but each contained a certain type of drawing, such as those showing underground geology. Within each drawer, every map had a four-digit number that tied it to the overall index card system. In the drawers, the drawings were arranged sequentially. Maps that were stored rolled often had filing information attached to the end. A typewritten list of the topical contents of Quincy's vault pigeonholes has survived, showing that the maps were filed according to coverage: "Old Miscellaneous," "Underground," "1st and 2nd Hillside Additions," and "Carpenter Shop, or Construction" were just a few of the numbered categories.[37] To make them easier to locate in pigeonholes, rolled maps could have a tag permanently attached to them (as we saw in figure 1.4). By contrast, Quincy used special metal clips that held together the rolls and provided the map number. In sum, as a mining organization grew, the infrastructure it used to store and retrieve visual tools such as maps, drawings, and blueprints needed to change as well.

Underground surveying and mapping has not always been part of mining. Georgius Agricola, German author of the ur-text of mining, *De Re Metallica*, reported simple surveying of the underground in 1556, but these surveys did not become maps. "There are many arts and sciences of which a miner should not be ignorant. . . . Fourthly, there is the science of Surveying that he may be able to estimate how deep a shaft should be sunk to reach the tunnel which is being driven to it, and to determine the limits and boundaries in these workings, especially in depth."[38] As can be seen in figure 1.5, Agricola's surveyor used cords to create a series of corresponding right triangles, permitting him to measure angles and distances, though they did not express angles, in particular, numerically. However, Agricola's surveyors did not map their mines in a modern sense. Instead, they platted their surveys on a surveyor's field—a full-scale, open-air, temporary reproduction on nearby level ground. These surveys were chiefly used to ensure that tunnels and shafts being driven toward each other would meet, or that adjacent mines did not encroach on one another's property.[39] The earliest underground maps, according to mining engineer Herbert Hoover (who translated the work into English in partnership with his wife, linguist Lou Henry Hoover), were not made until after Agricola's time.[40]

With the development of the hanging compass for surveying in Europe

Figure 1.5. A surveyor depicted in the mining book *De Re Metallica,* first published in 1556. Such surveyors measured triangles and reproduced their results in full scale on the ground in an open-air "surveyor's field." From Georgius Agricola, *De Re Metallica,* translated by Herbert Clark Hoover and Lou Henry Hoover (London: Mining Magazine, 1912), 131.

after the mid-seventeenth century, it became easier to use the same instrument for making underground maps. However, this was apparently rarely done, and the maps that might have been created would not have been particularly accurate.[41] By the middle of the nineteenth century, mine mapping using optical instruments was taught at the leading European mining schools, but such mapping, like trained engineers, was scarce in America. Mining engineer Edwin J. Hulburt noted that in the early years of mining in the Michigan copper country in the mid to late 1840s, when Cornish influence was strong, the crude underground surveys made with the compass were still reproduced at full scale on the surface.[42]

American mining accelerated dramatically with the discovery of gold in California in 1848, but underground mapping did not automatically follow suit. Mining in California during the Gold Rush era was placer mining, conducted close to the surface. Exploiting the free gold of placer claims required water, gold-bearing dirt, and washing technology that ranged in sophistication and scale from simple pans and rockers to long toms and elaborate sluice boxes. Even at their most complex, however, these operations had little need for underground maps or for mining engineers. Gold Rush observer John S. Hittel described placer mining companies that would pair with their neighbors to create a joint tunnel, dug along the property line between the enterprises, to access pay dirt. "Once having reached the pay-dirt, a surveyor is called, and he marks out the line of their claims, and a drift is usually run by the compass along the dividing line . . . from that time forward, each company knows where its dirt lies; each keeping on its own side of the drift."[43] In these tunnels of the 1850s, no map was required, because the tunnel itself served as the property line. Similarly, the quartz miners' regulations for both Nevada County and Sacramento County, in California, which were a sort of governmental compact regulating the claiming and working of mines in those areas dating from the 1850s, mentioned maps only to allow a miner to have one made and file it along with his description of a claim, "if deemed requisite to more particularly fix the locality."[44]

The use of underground maps was closely associated with the industrialization of mining in the late nineteenth century, as mentioned in the introduction. When mining practices were relatively simple, there was little need for mine maps, underground surveying, or mining engineers. Early mining in the United States—that is, before the mid-1850s—was generally a localized affair. Mines were relatively shallow, and the minerals that preindustrial miners sought rarely needed much complex processing after their extraction from the earth. Accumulated local knowledge generally provided the basis for mining practice, and some foreign experts, many of whom were skilled miners from Europe, supplemented local methods. Occasionally a graduate of a foreign mining school might be hired as an adviser, but mining in pre-1850 America rarely posed problems of a sort that required a college-trained engineer to solve, and maps were rarities at best. For example, one Pennsylvania anthracite mine, which started production in 1830, was unmapped for thirty-four years.[45] The hands-on knowledge of skilled miners in the European tradition was more valuable in this context for directing productive work in the anthracite collieries of Pennsylvania. The same held true for mining practices based on Mexican and Spanish antecedents. For example, in underground mining for mercury at New Almaden, California, in the 1850s and early 1860s, a contemporary observer noted, "[T]he business of searching for the large deposits is never intrusted [*sic*] to educated mining engineers, but always to mining captains, who have themselves been laborers, and have learned by experience where to seek."[46] At best, a mining engineer may have been regarded as excess baggage. Edwin Hulburt recalled the attitude of the skilled Cornish miners in Michigan's copper country before the Civil War: "[I]n fact, the necessity for the skill of a mining engineer was looked upon askance, and generally he was regarded by the miners as a genteel supernumerary."[47]

Mine mapping spread at an uneven rate through the American mining industry, beginning in the late 1850s and accelerating in the mid-1870s and 1880s, to widespread acceptance in the early twentieth century. In this pattern it reflects the bumpy history of the industrialization of mining in America, as important districts grew into prominence and frequently faded. Underground mapping might be advanced in certain districts more than others for reasons beyond simple technical expertise. For example, the anthracite coal mining industry of eastern Pennsylvania, which we will investigate in greater detail in chapter 2, was an early leader in American mine surveying due to a host of factors, including early state regulation requiring maps to be made. In the 1880s some other states passed mine safety laws that required mapping, which undoubtedly over time increased the number of mapped mines, though the spread of such legislation to all states took until the first decades of the twentieth century. By contrast, some outdated practices, such as surveying with the hanging compass, persisted in parts of the bituminous coal industry into the 1890s.[48]

In smaller mines earlier in the nineteenth century, the whole process of surveying and constructing the map might be done solely by a mining engineer, utilizing mine laborers on special duty to help with the surveying tasks underground. In many cases such an engineer was not a regular employee of the mining company, but was hired from a pool of locally available engineering talent. For example, I. E. James on the Comstock Lode and T. S. McNair in the Hazleton anthracite area both served as mapmakers for many mines, though they occasionally held regular positions as engineers as well.[49]

This pattern of consulting engineers remained strong as long as the visual culture of mining was not tightly integrated into a mining company's operations. In a broad sense, larger operations were the first to acquire in-house mapping capabilities. Many smaller firms still contracted out their engineering expertise into the twentieth century, because of the expense of engineering work and because the company was not dependent on the visual information a well-made mining map might provide. "A poor map is often the result of the operator's idea that if a contracting engineer visits the mine only once or twice each year for a couple of days and makes a map that will pass the inspector he possesses something that supplies his needs. The requirements of the present have been met, and the average cost per ton of coal for the year has been reduced by not having a regular force to look after the mapping of the mine. But the future has been neglected."[50]

Both the industrialization of the mining industry and the development of the visual culture of mining accelerated in the 1880s. By the end of that decade, accurate mine surveying was widely understood to be an important component of any mining engineer's work, though best practices were followed consistently only at the largest mining firms. The practice of underground mapping in America was most advanced in the anthracite region in Pennsylvania, due in part to the pioneering mine safety laws passed by the state that demanded the mapping of anthracite mines.[51] An editorial in *The Colliery Engineer* in May 1888 noted, "The profession of mining engineering has advanced more in the past decade than any other profession, and in that portion of the engineer's work, comprised under surveying and mapping, great advances have been made."[52] The editor praised the work of the larger anthracite companies in mine surveying and held them up as examples. He urged other engineers, especially bituminous coal engineers, to follow their example of careful mapping from the start to avoid problems created or exacerbated by unmapped mines. The editor made it clear that the goal was a "perfect" map, loaded with information, gathered by the observant (and hard-working) mining engineer, even as his veiled critique hinted at best practices commonly left unpracticed:

> The engineer should use his judgment and his mental faculties continually. With a modern mining transit, he should read the vertical as well as the horizontal angle at every sight. He should note every peculiarity or fault in the vein. He can not take too many notes regarding the inclination of the vein, nor can he make his notes too complete in any sense. Paper and pencils are cheap, and the engineer's note books should contain almost a perfect history and description of the development of a colliery: His maps should be a perfect picture of the ground plan of the inside working, outside improvements and topography of both the surface and the bed rock of the vein. Every point possible should have its elevation above a common basis marked. The date of every survey should be lettered at the places where the faces of the workings stood when each survey was made. Every synclinal and anticlinal axis, whether local or general, should be prominently shown. Each pillar robbed out should be shown as gone. The air courses should be prominently marked, and whenever practicable the course of the air through the workings should be marked by arrows; doors and brattices should be shown; and, in fact, every little detail possible should be distinctly noted and marked, for there is no telling at what moment the information concerning these points will be needed.[53]

By the 1880s some mining engineers and educators had begun to recognize the value of mine mapping, but apparently at least a significant minority of mining companies had not yet made the visual representations part of the fabric of everyday mine management. The legal requirement of several states to make maps was certainly perceived by smaller operators, in particular, as an unwanted financial and operational burden—one engineer characterized small operators as seeing required maps as "a parchment exacted by a tyrannical legislature."[54]

An exasperated Iowa mine inspector noted in 1888 how the state law requiring mine maps was "honored more in the breach than in the observance," in Wayne County coal mines. He used one long, breathless sentence to describe the run-around he received from area miners when he asked for their map: "[W]hen we visit a mine and inquire for the map the pit boss or man in charge will probably say that it is at the main office, or that we have no map of our mine, or that it is not finished yet, and will promise to have it ready when we visit them again, and it is not always convenient at the time to go to the main office."[55]

Experienced engineers made efforts in the technical press and government reports to promote the value of mine maps to their unenlightened brethren. Mining engineer B. W. Robinson's article advocating and explaining mine maps appeared in the *Report* of the Inspector of Mines of Kentucky for 1888, and was reprinted in the journal *Colliery Engineer* in 1889. Robinson claimed that most mine managers were "slow to realize their necessity, and hence have them attended to only when works are in close proximity to the limits of their property, or to satisfy the requirement of the mine law." These unenlightened managers, Robinson claimed, used "careless or incompetent persons" to finish the work cheaply, which often consisted merely of measuring distance traveled and not true surveying.[56] Inaccurate maps could give a general sense of the mine at best, and would be useless for understanding what parts of the surface were undermined. Old, unmapped workings posed hazards to miners from floods, built-up gases, and spontaneous fire in waste materials.

"A good mine map will many times repay the operator for the outlay attached to it," argued Robinson, for it could be used to estimate materials needed, keep track of coal mined already, help engineers route new tunnels for maximum efficiency, and help solve problems of ventilation and water drainage. By using different-colored inks, mining engineers could ensure that the support pillars in multiple-level mines lined up. "In fact, no mine can be altogether successful, which is not properly represented by a good, accurate map." Robinson suggested beginning surveys at the moment the mine was opened, by establishing permanent surface base lines to orient all future surveys. Using a transit—a magnetic compass was not suitable for mine work because of the amount of iron in the form of rails and pipe typically found underground, warned Robinson—the mine should be surveyed at a maximum interval of six months; any longer, and portions of the mine, "frequently those which are most important, become inaccessible."[57]

To mining engineer Robinson, "the prime object of a mine map is to show the underground workings in their relation to the boundaries of the company's property, the amount of surface undermined and what remains, and to serve as a complete record of all the workings." While underground surveying had been used to help respect property lines since Agricola's time, the notion of a mine map as a "complete record" was rather new, and pointed toward the future of mine maps as information storage technologies. Since there was "no telling when such information may be needed," Robinson counseled his readers to keep "very volumi- nous" survey notes. "The engineer should not forget anything that would be of use in making his map, or for future reference," suggested Robin- son. He advocated keeping sketches of the mine along with the survey notes, but Robinson's wide-ranging view of information saturation was expressed best by his suggestions for what to include on mine maps. If a mining engineer followed his directions, his map would contain an air line, surveyor's stations with connecting lines, tinted working stations, all the property lines, roads, streams, outcrops, "general direction of hills and valleys," permanent objects or surface landmarks, underground elevations (though only of "prominent points"), and direction and measurement of the dip of the vein. In addition, Robinson recommended keeping a set of profiles (or sections) of the "principal entries," with information about grade and elevation of haulage ways as well as notes on the height and "character" of the tops of tunnels.[58]

In sum, by the late 1880s at least some mining engineers, such as B. W. Robinson, recognized the potential power of underground maps beyond mere legal compliance and avoidance of trespass. Instead, Robinson under- stood maps to be complex objects that organized many different types of information spatially; once created, this document served the engineer as a visual index of other information and as a tool that, through its concate- nation of different types of data, helped engineers visualize solutions to the problems of mining.

Decades later, the biggest coal mining companies had embraced the potential of mine maps, but smaller operators still frequently saw maps as a noisome regulation to be complied with by expending a minimum of effort, rather than as an integral part of mining operations. In 1915 min- ing engineer H. L. Fickett attempted to persuade smaller operators, who he believed were skeptical of the value of mine maps, by presenting short case studies of operational problems that could have been solved by having good underground maps. In one case a mining company that had updated its map carelessly and infrequently over a period of years found it "not worth the paper it was printed on" when a new haulage road needed to be constructed, and as a result the whole area had to be resurveyed. Fickett's second example pointed to a dispute between a mining company and a landowner, where after the mine was closed the landowner believed the mining company had underpaid royalties. The company's attorneys did not realize the map was inaccurate, and thereby agreed that it could serve as a basis for the royalty determination. The map indicated that some areas had been mined despite the fact that in actuality the conditions in those places had prevented the completion of the mining once it commenced. As a result, the mining company "paid out thousands of dollars in royalty for coal that was never mined . . . but which was shown as extracted by a mere sketch that bore the dignified title of 'mine map.' "[59]

By the first decades of the twentieth century, mine maps had become an indispensable part of the regular work of large mining companies. Those large companies had developed a visual culture of mining that permitted them greater coordination and control of underground spaces, labor, and the cost of production. Furthermore, the creation of maps had become a point of law for coal mines, as every coal mining state had by 1915 enacted laws requiring companies to make maps and furnish them to the state.[60] Clearly, mapping and surveying were considered part of the common duties of mining engineers across the industry. In a 1916 advice book targeted at young engineers, F. B. Richards described mapmaking as part of the regular duties of mining engineers. "As the mine is operated, accurate maps must be kept up of all the underground working [sic], which means much underground surveying and work in the drafting room if the property be a large one."[61]

Even though the practice of mine mapping had become widespread and a key part of the duties of mining engineers, variations in mapping practice between mines and districts was still common. "It has been my experience in visiting the drafting rooms of various coal-mining companies that invariably I have found everyone, from chief draftsman to blueprint boy, curious to know the methods of surveying and mapping employed by the company I represent. Also, they have seized upon small details that I had supposed were known to everyone and applied them with considerable advantage to their own particular system," noted mining engineer W. L. Owens in 1917.[62] Similarly, there remained a spectrum of accuracy and care with which the maps were created, likely reflecting the organizational commitment to the visual culture of mining within a firm. Managers and engineers at the largest firms had embraced the power afforded them by the visual culture of mining, but some lower-end operations did only the minimum mapping required: "[I]t is a well-known fact that many of the maps submitted are in reality nothing but mere sketches, beautiful from the viewpoint of the draftsman, but owing to abounding inaccuracies practically worthless to the operators. In such sketches the letter of the law has been fulfilled, but the spirit has been neglected." Fickett placed the blame on the small operators, who frequently made short-sighted decisions for economic reasons. The engineer hoped that the small operators would realize, "to neglect or slight [their] maps is not economy, but extravagance in the long run."[63]

By the twentieth century, mining engineers came to mining companies already knowing how to make and use underground maps, because they learned it in their university training. Mining engineering education became an important site of the development of a visual culture of mining.

As mentioned above, before the late 1850s few mining engineers were needed in the American mining industry. As mines were pushed deeper and ore-processing problems became more complex in the 1850s and 1860s, the utility of trained engineers increased. These included foreign-trained engineers; American students who graduated from foreign mining schools; engineers with related training received on the job from railroad, canal, or bridge construction; and graduates of American scientific schools. Additionally, in the period 1860–1920, some people who performed the work of engineers were not formally educated The formation of systematic state and federal geological surveys also provided a training ground for a generation of geologists and engineers who applied geological knowledge to the solution of underground problems.

The first American schools of mining engineering were formed in this context of the increasing use of engineering techniques to solve mining problems. Initial attempts in the late 1850s resulted in a handful of graduates. Before the 1862 passage of the Morrill Land Grant Act, which provided federal support for technical education through grants of federal lands that could be sold to provide permanent revenue, only eight universities in the United States (plus the two military academies) taught engineering in any form.[64] The Polytechnic College of Pennsylvania, a private university that closed its doors sometime after 1878, offered a track in metallurgy from its founding in 1853 and created a school of mines in 1857 that offered degrees in mining, making it the first American college to do so.[65] However, there were few graduates in mining fields and the school received little recognition. Though historians have probably underestimated the numbers of graduates of the school, it is unlikely that the Polytechnic College of Pennsylvania had much impact on the American mining engineering profession.[66]

The first truly successful American mining school was formed at Columbia College, in New York City, in 1864. Over the next decades, a number of schools started mining engineering programs, and most of the universities of states that featured major mineral industries formed mining schools, sometimes as separate campuses. The core curricula focused on geology, engineering, geometry, mathematics, chemistry, and physics, and almost always included practical instruction on surveying, drafting, blowpipe analysis, assaying, and the like. Most mining engineering programs required students to spend their summers in mines, gaining hands-on experience. Mine engineering, as taught in the classes and summer programs of late-nineteenth-century American colleges, included lots of sketching and drafting, a scientific approach to problem solving, and a preference for solid numbers. The overall aim of the schools was to unite engineering theory with mine practice, and to cultivate a wide-ranging vision of the mine as a controllable (or semicontrollable), rational system.[67]

Formal training in the techniques necessary to produce mine maps was instituted a few decades earlier in Europe, and was being taught by the late 1850s at some mining schools there. In the United States, only drawing, not cartography or surveying, was part of the mining curriculum of the School of Mines of the Polytechnic College of Pennsylvania.[68] Neither subject was included in the early curriculum of the Columbia College School of Mines, but surveying was later added as a topic to be learned during summer excursions to working mines.[69]

Marrying technical education with practical knowledge of the underground was essential for the success of mining engineering as a profession, since theory alone was not particularly helpful in an underground space that was contingent on a host of sensitive factors and prone to complex failures. Early scientifically trained engineers earned a bad reputation among mine owners for relying on theory that did not work under the specific conditions of a particular mine.[70] As a result, leaders of the mining engineering profession from the 1870s onward were careful to point out the need for graduates to gain practical experience. Eckley Coxe, in an 1878 presidential address to the members of the American Institute of Mining Engineers (AIME), lauded "the true engineer," who, according to Coxe, "mastered both the theory and the practice of his profession." This ideal mining engineer "neither neglect[ed] the theoretical speculations of the chemist, the physicist, the geologist, and the mechanical engineer, nor the facts and opinions of the practical men whose daily contact with the actual working of the furnaces enables them to make very valuable suggestions."[71]

Eckley Coxe was an anthracite mining engineer educated in European mining schools whose operation will figure largely in chapter 2 of this book. Coxe also played a direct role in creating the pioneering summer course for the students of the Columbia College School of Mines. He responded enthusiastically to the suggestion of Henry S. Munroe–who at the time was adjunct instructor in mining engineering at the Columbia College School of Mines–that mining engineering students should gain practical experience by working in the mines during the summer. Coxe invited the first group to his collieries at Drifton, where thirteen students spent July and August 1877 learning from the miners and engineers. Underground surveying was explicitly mentioned as part of the material to be learned, and the students were also required to sketch extensively and work on a detailed report. The experiment in Coxe's mines in 1877 was so successful that the trustees of the college voted to make summer trips a permanent part of the mining curriculum.[72] Later, the young engineers who experienced this summer program, such as Thomas H. Leggett, recalled the trip as being "of immense benefit."[73]

In the mid-1880s, instruction on the techniques of mine mapping became more common. By 1886, underground surveying was taught theoretically and practically at most of the major mining schools. An examination of the top mining schools of 1886 found that Columbia College, Lafayette College, Lehigh University, University of California, University of Illinois, University of Pennsylvania, and Washington University gave both classroom and field instruction in underground surveying. The University of Wisconsin and the Massachusetts Institute of Technology provided only classroom instruction in underground surveying, and the University of Michigan mining engineering program did not address the topic of underground surveying at all.[74]

Technical literature describing mine surveying was also rare before the 1880s. Prior to the late 1850s, only a handful of treatises about the topic had been published, many of them in German and Latin; those few written in English became quite rare over the intervening decades.[75] The topic was apparently new enough in 1883 for Lehigh University to grant an engineering degree to George Hood for a thesis on the subject, which reviewed a handful of articles in the technical press and described practical experience Hood had collected at a nearby coal mine.[76] Toward the end of the 1880s, however, the situation changed. Bennett H. Brough's popular text on mine surveying appeared in 1888, and went through many subsequent editions.[77] From the 1890s, a multitude of books covered the topic of mine surveying, and general purpose guides about surveying and about mining engineering generally carried at least a chapter or two about mine surveying.[78]

Underground surveying and mapping not only helped mining engineers better understand and control their mines, but also contributed to their professional identity. The power to make and wield technological representations set mining engineers, as a professional group, apart from others who could purport to do similar work, such as so-called practical men. In an 1888 article, bituminous coal mining engineer Reuben Street exhorted the practical, professional, and even cultural value of mine surveying for his fellow engineers: "The study of the principles of mine surveying will enable us to reach a point far above that which we now occupy, and to prove that the coal mining fraternity has in its ranks men whose minds are as susceptible to and capable of development, even in the higher arts and sciences, as are found among the men of other avocations."[79]

By the same logic, a poorly made mine map was a mark of a bad engineer. Mining engineers should "take pride" in creating accurate underground maps, argued H. L. Fickett, and this responsibility extended to using up-to-date equipment and techniques, such as using optical transits instead of magnetic compasses and recording both distances and bearings for all work. "The engineer should devote as much of his time to making a good map as to executing a piece of construction that is to serve the company for years to come," posited Fickett.[80]

Further evidence of the close tie between mining engineers and the maps they made might be found in occasional mention of the emotional investment that engineers made in their maps. Engineer George L. Yaste opined, because "an engineer generally regards his maps as a thrifty housewife does her articles of furniture," to see company officials treat the maps incautiously was "disheartening" and "discouraging." After all, the engineer had mothered the map into existence, and an emotional attachment to the visual artifact was understandable. "[The mining engineer] has seen it grow under his hands from a blank piece of paper to the valuable map it now is. He regards it as something of his own—a part of himself." With such an emotional connection to the mine map, it is understandable that a mining engineer witnessing "the ruthless manner in which [his map] is maltreated" would result in a "demoralizing influence." After all, the engineer's professional identity, not just his work product, was at stake.[81]

The classic symbols of the mining work hierarchy support this close connection between mapping and mining engineering. Miners were represented with pick, sledge, and shovel. Foremen, or "bosses," claimed the symbol of the safety lamp, which represented their power to inspect for explosive gas and prohibit or permit the miner to work. Engineers were represented with transit and notebook—the tools of surveying and constructing technological representations, key components of the visual culture of mining.[82]

In January 1877 the engineering and mapping team for one district of the Philadelphia and Reading Coal and Iron Company's (P&R C&I) numerous anthracite coal mining operations posed in a studio for a group portrait. This photograph (figure 1.6) vividly shows the importance of participating in the visual culture of mining to the professional identity of mining engineers. As was common portraiture practice at the time, the men were pictured with objects that symbolized their identity, posing in this case with the surveying tools that they used to create underground mine maps. The mining transit stands high on its tripod in the center of the photograph, almost like an independent member of the crew. The man in the center deploys the survey crew's notebook, while the man to his side holds surveying tools in both hands. The man seated below the transit cradles the target that the other engineers would see through the

Figure 1.6. Mining engineers from the Philadelphia and Reading Coal and Iron Company posed with their engineering tools for a group portrait in 1877, showing how the creation of visual representations had become intertwined with their professional identity. Hoffman Photograph Collection, Division of Work and Industry, National Museum of American History, Smithsonian Institution.

telescope, and beside him his colleague holds a folded-up surveyors chain. All of the men are identified as underground mining engineers by the teapot-shaped oil lamps they wear on their hats. This detail was of such importance that the photographer went to the trouble of smudging the negative to make it appear as though the lamps are lit. These men shared the symbol of the underground, the cap lamp, with other mine workers, and might share some of the surveying tools with ordinary civil engineers, but the two together marked them as mining engineers.

In conclusion, maps of underground mines were a central part of the visual culture of mining that developed in the late nineteenth– and early twentieth–century mining industry. Irrelevant or nonexistent before the late 1850s, underground maps gradually became more sophisticated, accurate, and useful, eventually becoming a key tool for mining engineers and the managers they served. To make a mine map, engineers had to use specialized tools to survey the underground, then draw, store, and duplicate the map in a space dedicated to the visual culture of the mine. The close association of mining engineers with underground mapping began, for most, with their collegiate training, if educated in the burgeoning mining engineering programs of the 1880s, or with exposure to the topic in technical treatises that emerged at the same time. Whatever the source, mine mapping and surveying became central to the professional identity of mining engineers. Creating and using maps made mining engineers who they were.

The next two chapters will examine two histories of mine mapping in further detail. Chapter 2 examines the anthracite coal fields of eastern Pennsylvania, where a series of factors made the underground maps of the anthracite region among the most accurate anywhere, and the anthracite engineers leaders in the emerging mining engineering profession. Underground anthracite mine maps themselves changed over time, and the addition of new information changed the uses of them.

Chapter 3 moves forward and westward to the copper mines of Butte, Montana. There, at the turn of the twentieth century, mining engineer David W. Brunton and mine geologist Horace V. Winchell reimagined the traditional engineer's mine map, changing its forms and rethinking its possibilities. Though their efforts grew out of the need to shore up their company against substantial legal threats, the system they developed would form, within less than two decades, a new model for how companies mobilized the visual culture of mining to see underground.

chapter 2

Anthracite Mapping and
Eckley Coxe

This chapter moves from the broad focus of the previous chapter to examine the visual culture of mining in one particularly important sector of the nineteenth century mining industry, the anthracite coal fields of eastern Pennsylvania. As shown in chapter 1, after the mid-1800s mapping gradually became an essential component of every facet of underground operations, but such was not the case before that time. The mining engineers of Pennsylvania's anthracite coal operations, with assistance from a pioneering safety law, played a large role in creating and popularizing the tools and techniques of extensive, accurate underground mapmaking. Mapping became an important responsibility of mining engineers, and a marker of their professional identity, as they competed with other occupational groups for jurisdiction over the underground landscapes of the anthracite fields. The anthracite maps themselves, as a visual technology that both represented underground geographies and helped create those spaces, changed significantly over time. These changes in underground maps both represented their increased use by engineers and managers to understand their mines, and also gradually made the maps more useful and thus a more important part of mining operations.

A combination of factors—including talented scientists and engineers, a tradition of scientific practice, exploitable resources, unintended consequences of legislation, nearby instrument builders, and some fortuitous timing—made Pennsylvania's anthracite region a site of underground mapping innovation. More-accurate maps enabled the anthracite companies to locate and extract coal more easily, plan production more systematically, and watch carefully over royalty payments—though not all of them did at first.

One of the most innovative and well-respected mining engineers in the anthracite mining business in the late nineteenth century was Eckley B. Coxe, whose mining maps from Drifton, Pennsylvania, illustrate larger trends in the changing underground mapping practice of the late nineteenth century. Coxe was born into a wealthy Philadelphia family. His grandfather, Tench Coxe, assembled the vast tracts of coal-bearing lands that his heirs, including Eckley, would later develop. Eckley Coxe studied at the University of Pennsylvania, then spent most of the Civil War years studying mining engineering at the Bergakademie in Freiberg, Germany, and the École des Mines in Paris. Upon his return to the United States in 1865, he (along with family partners) set up coal mining operations at Drifton on lands leased from his grandfather's estate. Gradually, Eckley Coxe and his partners consolidated their control of the estate's coal lands.[1] Their company was one of the largest independent producers of anthracite during its best years in the 1880s and 1890s.[2] Coxe played an important role in the professionalization of mining engineering by encouraging the development of institutions and periodicals, by supporting engineering programs, and by pushing for greater application of science and mathematics to mining activities. In short, Coxe's life is intertwined with the story of the professionalization of mining engineering and the development of the visual culture of mining, which makes his operation a compelling case study of the development of underground mapping. This chapter first describes anthracite collieries, like those designed by Eckley Coxe, which will set the stage for a discussion of underground mapping and mining engineering.[3] A series of maps from one of Coxe's anthracite mines follows; we examine them to see how the maps, and the context of their use and usefulness, changed over time.

Anthracite, or "hard coal" mines are found in North America almost exclusively in northeast Pennsylvania. (The presence of anthracite is in distinct contrast to bituminous or "soft" coal, which was mined in commercial quantities across the United States, in the Appalachian region, the Midwest, Colorado, California, New Mexico, Texas, and Alabama, among other places.) Anthracite contains extremely high carbon content and few volatile organic compounds, which means that it is difficult to ignite but once lit burns very hot with a clean flame and little smoke, making it a preferred choice for domestic heating and cooking. Anthracite collieries varied considerably in their details, but it is possible to give a general description of these work sites. Anthracite coal came out of the mine, was sized and processed in a factory-like building called a breaker located nearby, then usually was loaded into railroad cars for shipment to urban markets. The initial step of this chain of production, distribution, and use took place in a subterranean landscape spatially organized by a mining engineer for the purpose of facilitating the production of maximum quantities of anthracite coal at minimum cost. It was this space that a mine map described, and it was likewise this space that mining engineers tried to control with the use of maps.

The shapes and sizes of the anthracite mines were strongly influenced by the geology of anthracite coal. In eastern Pennsylvania, several seams or veins of anthracite exist atop one another with intervening strata of slate, shale, and sandstone. The strata that contain the coal were subject to enormous geological pressures, which both transformed the coal into anthracite, and folded and buckled the layers of strata. As a result, Pennsylvania anthracite veins can pitch quite steeply and unpredictably, and outcrop on the surface in many places. Figure 2.1 illustrates a vertical section through Eckley Coxe's mines at Drifton, as the seams were understood to exist in the late nineteenth century; it shows clearly the folded and buckled parallel coal seams.

Commercial mining of anthracite in Pennsylvania began in the 1820s and 1830s, but for decades the industry had little need for underground maps. In the early years, Pennsylvania anthracite miners simply followed the coal downward in an open pit, a process akin to quarrying. By the 1850s, miners were sinking shallow shafts, or slopes, in addition to the pits, but underground mapping was still almost nonexistent.[4] Beginning in the 1850s, the physical plant of these operations expanded dramatically, both above- and underground, and its elements continued to be refined as long as underground operations were ongoing.

First, some means of reaching the coal underground were needed. If the coal was in a hill, above the waterline, a tunnel driven almost horizontally would reach the seam and provide an easy way in and out of the mine. If the coal was below ground, engineers could drive an inclined tunnel

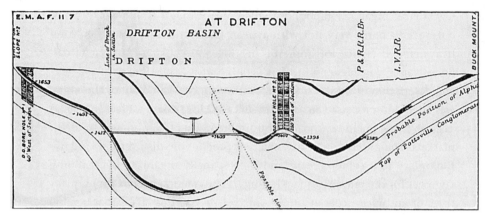

Figure 2.1. This vertical section shows the coal veins under Eckley Coxe's mines at Drifton, Pennsylvania, as they were understood to exist in the late nineteenth century. Anthracite veins dipped, rolled, and undulated, adding to the challenges of exploration, mining, and surveying. From Benjamin Smith Lyman, "Folds and Faults in Pennsylvania Anthracite-Beds," *Transactions of the American Institute of Mining Engineers* 25 (1896): 343.

along the angle of the coal seam. This inclined tunnel was called a slope. Slopes were generally favored in the anthracite region during the nineteenth century for several reasons. Since they were dug through the coal, it was much easier to make sure that the vein was not lost due to a misunderstanding of geology. Additionally, the anthracite was softer and easier to tunnel through than was the surrounding rock. The extra coal gained from excavating the slope was of only minor importance. It was also possible to dig a vertical shaft to intersect the coal seam at a predetermined point. The collieries of Eckley Coxe at Drifton used slopes, which followed the Buck Mountain coal seam underground at an angle of twenty-five to thirty degrees.[5]

Once the slope, shaft, or tunnel reached a convenient point, which often was the bottom of the coal seam (the syncline), miners drove horizontal tunnels called gangways, branching off in both directions along the plane of the coal seam. The gangways (together with the slope) constituted the main transportation and ventilation infrastructure of the mine. From the gangways, miners constructed a series of working chambers called breasts to remove the coal. Each breast was separated from the next by a wall or pillar of coal left in place for support. When the coal seam was tilted, miners made these breasts uphill from the gangway, which made it easier to load coal into cars with the help of gravity. A miner and his helper were assigned to an individual breast, the working of which might take a year or more. The miner's job was to direct the working of the breast, bore holes in the face, load them with powder, and blast down the coal. The helper then loaded the coal into a mine car, which a mule driver then hauled to the foot of the slope, where the car was then hoisted to the surface.[6]

The surface plant of a colliery was dominated, after the 1850s, by the breaker, a large building housing coal processing machinery. Coxe's Drifton No. 2 breaker circa 1891 is shown in figure 2.2. The breaker commonly contained the hoisting machinery as well. After coal was hoisted to the top of the breaker, it was fed by gravity through a series of spiked rollers that broke the coal into useful sizes. The broken coal then passed through rotating or gyrating screens, which sorted the anthracite into standard market sizes. In some breakers, including the iron breaker Eckley Coxe built at Drifton, the smaller sizes of coal passed through an automated washing machine, which separated coal from slate on the basis of specific gravity. Washed or not, the coal slid in troughs past employees who manually picked pieces of slate from the anthracite. These employees were usually children, called breaker boys, as well as disabled miners. After being cleaned by the breaker boys, the coal slid into the appropriate pocket or storage bin, which could be emptied at the bottom into a railroad gondola. The waste generated in the process of preparing coal for market consisted of slate and other rock, which was dumped on a slate pile, and very small coal, coal dust, and dirt, which was known as culm and dumped on a culm pile.[7] The broken and sized coal was then hauled to the major markets, initially via canal but in later years by railroad; from shipping points such as New York City it was sent to customers across the country.

As mentioned in chapter 1, the American mining industry began to change in the years surrounding the Civil War. Efforts were made in the late 1850s to start formal education in mining engineering in America, first in Philadelphia (an experiment that failed), then in New York City (one that succeeded). Similarly, technical journals aimed at a nationwide audience of mining professionals began—and sometimes ceased—publication starting in the 1850s, with more-successful efforts stemming from the 1860s. In other words, mining engineering began to create for itself many of the institutions that identified it as a profession, including

Figure 2.2. The "Iron Breaker" at Drifton No. 2, built by Eckley Coxe, circa 1891. To the upper right a conveyor deposits dirt and coal dust on a growing culm pile. Coal was hoisted from the mine along the ascending entrance to the top of the breaker, where it was washed, crushed to size, and picked clean by a workforce of "breaker boys," some of whom posed here in front of the enormous building. Scanned from the negative by Eric Nystrom (2005), Division of Work and Industry, National Museum of American History, Smithsonian Institution.

common practices based on the scientific method and shared technical information, educational institutions, and technical journals. Many of these efforts were led by American mining engineers who received formal engineering training overseas at long-established, respected institutions such as the Bergakademie in Freiberg, Saxony; the École des Mines in Paris; the mining academy in Clausthal, in present-day Austria; and other smaller institutions.

Pennsylvania's anthracite mines certainly benefited from this influx of professional talent. The anthracite industry's close alignment with the burgeoning Pennsylvania iron industry, which used (in places such as Bethlehem, for example) abundant clean-burning anthracite coal to fuel blast furnaces, led Pennsylvania's anthracite region to become one of the nation's premier mining areas. Additionally, the region undoubtedly prospered, in part, because of its proximity to both Philadelphia and New York City, two cities that not only served as markets for coal and sources of capital, but also served as major centers for scientific and technical expertise. In turn, this nucleus of well-trained and enthusiastic engineering experts ensured that the anthracite region was well represented in further efforts to develop the mining engineering profession.

These experts took one of the most important steps of all, the creation of a national professional society dedicated to mining engineering (the second of the major American engineering professional societies to be created, predated by only the civil engineers). In 1871 three anthracite mining engineers—Eckley B. Coxe, Richard P. Rothwell, and Martin Coryell—issued a circular calling all mining engineers to meet at Wilkes-Barre in May for the purposes of organizing the American Institute of Mining Engineers (AIME). Of the twenty-two attendees at that first meeting, all but six hailed from the anthracite region.[8]

Even as the mining engineers met to form the AIME in 1871, the legal environment in which anthracite mining took place was changing, in ways that would ultimately benefit mining engineers. In 1869 the Pennsylvania legislature passed a pioneering mine safety law—the first in America—but due to serious disputes over its terms, necessity, and scope, it was limited in its provisions and applied only to Schuylkill County. Within a few months, a disastrous fire at the Avondale Colliery in Luzerne County killed 110 men and boys. In response, state lawmakers passed a more expansive 1870 mine safety law that applied to all of the anthracite fields, not just those in Schuylkill County. Both laws required that underground mines be surveyed and that the maps be submitted to the newly created office of state mine inspector on a twice-yearly basis.[9] In the context of the furor over regulation of any sort of the coal mines, the mapping provisions of the safety law seemed only a small part of the act. However, this legal requirement was eventually to help contribute to making the anthracite region a leader in mine mapping techniques.

A shift in mining practice of that magnitude, from frequently unmapped to almost universally mapped, did not occur evenly or without some difficulty, and the rise of mapping in the anthracite coal fields can shed some light on the process. By the late 1880s the mine safety law had contributed to making the Pennsylvania anthracite region a hotbed of mine surveying innovation.[10] Bennett Brough, addressing an audience primarily made up of British mining engineers, noted in the preface to his book on underground surveying, "Few mine-surveyors in Great Britain appear to be acquainted with the methods and instruments used abroad. This is the more to be regretted, as no mine-surveys made in this country approach in accuracy those of the collieries of Pennsylvania, or those of the metalliferous mines of the Harz."[11]

Charles A. Ashburner, who headed the Second Geological Survey of Pennsylvania's work in the anthracite regions, acknowledged the precision of the mapping efforts and even compared them to the extraordinarily precise American geodetic surveys: "Next to the surveys made by the United States Coast Survey, the mine surveys in the anthracite region are without doubt the most accurate of any of the extended surveys which have been made in America. I do not make this assertion without a thorough knowledge and appreciation of the other classes of work which have been done by the Government and by individuals in the United States."[12]

By the time of the passage of the 1870 safety law, mine maps were not unheard-of, but were not a standard part of each mine operation, either. For example, the anthracite mine at Beaver Meadow began production in 1830, but was not mapped until 1864.[13] The 1869 law, which applied only to Schuylkill County, required that mines be mapped, gave inspectors the right to examine the maps, and mandated that a copy of the map be furnished to the inspector when a mine (or section of a mine) was to be abandoned. However, the 1869 law did not detail particular standards the maps needed to meet.[14] The 1870 law, which applied to all Pennsylvania anthracite mines, increased the mapping regulations. Its first section required that each mine be accurately mapped; if there was no map, the operator had to make one. (The law gave the district mine inspector the right to hire someone to map the mines of owners who neglected or refused to map on their own, and charge the cost to the company.) This underground map was specified to be on a scale of one hundred feet to an inch, and had to show the workings, give a sense of the "general inclination of the strata," and the boundary lines of the company's land. This map (or a copy) had to be filed with the state mine inspector for the district and updated twice a year—far greater involvement than the 1869 law's simpler provision. As with the earlier law, if a mine or a portion of it was to be abandoned (or have the pillars robbed), the operator had to file a map and give notice to the inspector. Updates could consist of a new map or a report of recent progress.[15] Despite the ambitious requirements, neither law provided any penalty for noncompliance.

When the law went into effect, it was clear that underground mapping was practiced in some parts of the anthracite fields, but was by no means widespread. Statistics from the early annual reports of the anthracite mine inspectors, while uneven in their coverage and clarity, can help illuminate the halting steps toward comprehensive underground mapping. In the first report of the Pennsylvania mine inspectors, covering the initial year of enforcement in 1870, Schuylkill District clerk P. F. McAndrew named only 68 collieries that had provided copies of maps to the inspectors by the end of the year, of 202 active collieries in his district. As a result, even though the mines in this area should have had a head-start on the mapping requirements due to the 1869 law and the several months of leeway they had to come into compliance after the 1870 law passed, only 33.67 percent of Schuylkill-area collieries were mapped.[16]

The situation by the end of the year was a little better in the other anthracite districts, though the transition to mapped mines was not without difficulty.[17] In the Middle District, mine owners resisted all elements of the law, including but not limited to the mapping provisions. Inspector T. M. Williams took office after the law had been in force for months, but found that eleven of the twenty shafts in his jurisdiction had only one entry and had taken few or no steps to build the legally required second exit. In August, Williams served notice to comply or shut down. Most of the operators met in private to discuss the issue, and must have decided that operating within the law was the best course of action. "Soon after this surveys were made by most of them, and they began in good earnest to work according to law," reported Williams.[18] By the end of 1870, Williams was happy to note the "great deal [that] has been done in preparing plans and surveys of the mines," noting that in only "about a dozen instances" had he not received maps by the end of the year.[19] If each "opening" was considered to be a mine by inspector Williams, then the efforts of the mines to catch up to the law meant that 83 percent of mines in his district were mapped by year's end.[20]

Even with underground mapmaking enshrined into law, it was clear that it was not widely considered an important part of the overall operation of the mine. A short essay titled "Government of Mines" in the *Anthracite Inspectors Reports for 1871* described in detail all of the duties of a diligent mine operator, including taking care of matters of ventilation, accounting, and using steam engines, but never mentioned surveying or mapmaking as a goal in itself or a detail to be attended to.[21]

The compliance with mine map requirements increased slightly the following year, but was still not a given throughout the Schuylkill region. In 1871 inspectors reported 213 active collieries and 32 partially active collieries total in the three districts of the Schuylkill region.[22] They noted that a total of 102 collieries had furnished maps for the year ending December 1871, though some collieries provided more than one map when they had multiple shafts or slopes in operation. Additionally, some sixteen maps from Shamokin-area collieries had been provided in 1870, but were given to the companies for updates and never returned to the inspector. If we use only the figure for active collieries, 47.89 percent of Schuylkill collieries were mapped by the end of 1871.[23] Though using different figures might vary the computed compliance rate, overall it is clear that underground mapping was by no means a universal practice, even two years after the enactment of laws requiring it.

The *Anthracite Inspectors Reports for 1873* saw another increase in compliance with the law requiring mapping. Though the report gives slightly contradictory numbers for the total number of collieries in the Schuylkill District and the total number of maps on file, it is clear that the number of mapped collieries increased over the previous year, with 75–79 percent of district mines providing maps by the end of the year.[24] The tables that reported colliery ownership show a trend in 1873 that would soon result in expansion of underground mapping in the area. In 1873, the Philadelphia and Reading Coal and Iron Company (P&R C&I), a part of the Philadelphia and Reading Railroad, acted aggressively to bring Schuylkill mines under its control. By the time of the report, the P&R C&I was listed as landowner for a large proportion of district collieries, though the company exercised outright operational control in far fewer cases at that time.[25]

The following year saw no real improvement on the overall rate of underground mapping in the Schuylkill District, but more fine-grained information about map compliance allows us to see the regional variation even within the Schuylkill District. The mine inspectors' report of 1874 did not provide district-wide data on maps as before, but instead each of the three regional (district) inspectors provided the numbers. On the whole, by compiling the reports of the Pottsville, Ashland, and Shamokin areas, we see that in the Schuylkill District 127 out of 171 collieries provided maps in 1874, for a compliance rate of 74.27 percent. However, this average obscures varying rates of compliance within the district. The Pottsville region had significantly greater compliance than the district as a whole, with 42 of 48 (87.5 percent) collieries mapped. The Ashland and Shamokin regions, by contrast, had slightly less than 70 percent compliance, with 48 of 70 (68.57 percent) and 37 of 53 (69.81 percent), respectively.[26]

The number of anthracite collieries that were mapped increased significantly in the mid-1870s. During 1875, the anthracite mines were idled for nearly half the year due to a major strike, and perhaps this helped spur increased compliance with the mapping provisions of the mine safety law. Mining engineers in Pennsylvania and elsewhere frequently tried to conduct surveying work when the mines were otherwise idle, such as during strikes and on Sundays, because of the inconvenience working miners posed to surveying operations and vice-versa, and the engineers would not have been on strike. Additionally, as the P&R C&I extended operational control across much of the coal field, it likely spread advanced surveying techniques borrowed from or inspired by standard railroad engineering practice. (The mining engineering crew who posed in 1877 with their surveying tools, shown in chapter 1 figure 1.6, worked for the P&R C&I.) Though as usual some of the numbers might be a little off, it appears that of the 174 collieries in the district, 159 provided maps to the inspectors, for a compliance rate of 91.38 percent for the Schuylkill District overall. Most of the non-mapped collieries were in the Pottsville region, where the number of collieries submitting maps remained nearly the same as the previous year, while the overall number of collieries significantly increased.[27]

The ongoing problems with compliance with the mine mapping provision eventually prodded the Pennsylvania legislature into action. A bill was passed on May 8, 1876, that amended the 1870 act. The amendment made minor changes to clarify and expand on the details required for underground maps. The earlier act required boundary lines, but the amendment also forced owners to show the proximity of their workings to adjacent owners. However, the big change to the final clause finally gave the act some teeth to persuade owners to comply with the mapping requirement: "[I]n case the said owner or agent shall neglect or refuse to furnish the maps or plans by this section required, or any of them, or shall knowingly or designedly cause such maps or plans when furnished to be incorrect or false, such owner or agent thus offending shall be guilty of a misdemeanor, and upon conviction shall be punished by a fine not exceeding five hundred dollars or imprisonment not exceeding three months at the discretion of the court."[28] This new law, making it a crime not to provide accurate mine maps, may have persuaded most scofflaws in the anthracite region to finally map their underground mines. The mine inspectors' report for 1876 reflected a streamlined format that did not include a general list of Schuylkill District collieries that provided maps, perhaps because compliance had finally become extensive enough to be unremarkable. Inspector Samson Parton, in charge of the Pottsville region, was the only inspector to mention maps explicitly, and he did so to emphasize overall compliance in his district: "With the exception of a few new collieries all have furnished maps in conformity with the requirements of the law, and their extensions and corrections are properly attended to."[29]

While the mines in the Schuylkill District under Parton's supervision might have finally come into compliance, Inspector William S. Jones, whose inspection district to the north included the Scranton area, was having difficulty getting any maps at all: "Now, this clause in the law has been wholly disregarded, almost universally in this district, and I had, and am still having, what might be deemed unnecessary trouble with the corporations, in my efforts to have those maps furnished. There are miles of underground workings, of which there are no maps," grumbled the inspector.[30] Adding further insult, just because some mines had some kind of map by the mid-1870s did not necessarily imply that such a map was up to the high standards of the inspectors or useful to those not intimately familiar with the mine. They lacked survey stations, property boundaries, and measurements of the steepness of the coal strata, "all of which," argued Jones, "are important to have."[31] To Jones, a bad map was almost worse than no map at all. He took particular issue with the apparently widespread practice of underreporting mine work conducted before the passage of the law in 1870. Many operators would simply note "old workings" or "worked out" in areas that were closed before 1870.[32] Jones made the case that an underground map needed to include these areas as well to be truly useful in directing, for example, a rescue attempt after a cave-in. This possibility clearly struck close to home for Jones, as he related he had himself "experienced the painful sensation of being entombed in this manner" when a large cave-in occurred at Carbondale in 1846. Without maps at such an early date, Jones and some sixty others would have perished "had it not been that there was one man . . . who knew every yard of the workings, and who was brave enough to work his way through the old workings and over the edges of the cave."[33] According to Jones, an accurate map that included all of the workings of a mine, no matter how old, would be the only way modern rescuers could hope to reach trapped miners in time.

Thus, by the late 1870s, even after mine mapping had become a requirement in the anthracite country theoretically punishable by law, it was not a universal practice, and where done, was not necessarily done to the highest standards. This suggests that mine mapping was not yet an activity integral to the everyday work of mining for all firms. Within a decade, however, the promising start of precise mapping by requirement of law would transform the anthracite country's mapping practice into among the best in the profession, as noted above by Brough and the technical press.

One anthracite mining engineer and mine owner who helped lead the charge for professionalization of mining engineering was Eckley B. Coxe (figure 2.3).[34] Coxe was the driving force behind a family partnership, the Coxe Brothers & Co., that operated coal mines largely on lands leased from the estate of his grandfather, making Coxe recipient of both profits from the company and, to a lesser degree, of royalties as an heir to the estate.[35] Coxe was an early leader in the field of American mining

Figure 2.3. Eckley B. Coxe was a European-educated mining engineer and owner of several Pennsylvania anthracite mines, including those at Drifton. Coxe played an important role in encouraging the professionalization of mining engineering, helping to organize the AIME, serving as its president, and inviting mining engineering students to study his mines during their summer vacations. From R. W. Raymond, "Biographical Notice of Eckley B. Coxe," *Transactions of the American Institute of Mining Engineers* 25 (1896): 446–447.

engineering. He attempted to run his colliery at Drifton according to the best practices of the day, and was recognized for his efforts. A contemporary newspaper described Coxe's Drifton operation "the model mining plant of the world."[36]

Coxe was an innovator in mine surveying, which likely resulted from long exposure to the topic. The coal lands he managed had belonged to his family for more than a generation, and family trips included inspecting the coal properties. After graduating from the University of Pennsylvania, Coxe assisted Benjamin Smith Lyman in conducting a topographical survey of a portion of the family coal lands. While studying mining engineering abroad, Coxe worked with an early master of mine surveying, Julius Weisbach, who taught at the famous Bergakademie in Freiberg. Weisbach's contributions to the art of mine surveying in the first half of the nineteenth century included pioneering the use of the optical theodolite instead of a magnetic compass.[37] Coxe translated the German professor's thick calculus text into English.[38] Coxe must have read Weisbach's book (in German) about mine surveying, which was published in 1859, just a few years before Coxe's arrival in Freiberg.[39]

Coxe was a mining renaissance man of sorts—well-educated, scientifically and civically minded, driven, charming, and possessed of sufficient familial wealth to not be denied opportunities because of a lack of money. He participated enthusiastically in civic life, particularly in scientific and technical organizations. He helped to found the AIME in 1871 and held several offices, including two terms as president in 1878–1879.[40] Coxe was elected to the American Philosophical Society in 1870, and in 1894 he was appointed to the Geological Survey Commission of Pennsylvania.[41] He was vice president of Section D (Engineering) of the American Association for the Advancement of Science, and he served as president of the American Society of Mechanical Engineers. Coxe was especially concerned about waste and efficiency in energy consumption, spearheading a multidecade effort of the AIME to investigate the waste of coal in anthracite mining, and inventing and building a traveling grate furnace that made it possible to burn small sizes of anthracite coal that had heretofore been considered useless.[42] To support his scientific pursuits, Coxe amassed an enormous technical library at Drifton, containing both up-to-date and historical material, including a copy of Agricola's *De Re Metallica*.[43]

Eckley Coxe also was active in society beyond his scientific and technical interests. He was a committed Democrat, and was elected to the Pennsylvania State Senate in 1880. Famously, he refused to be sworn in to office because he would have had to affirm an oath that he had not spent money illegally for election. He publicly stated he laid out cash willingly and liberally, but not for any morally problematic reason; because he did not know the rules at the time, he could not, therefore, promise that all expenditures were legal. In the special election held in 1881 to fill the seat, Coxe was elected again with a far greater margin than before.[44] Coxe served as a trustee of Lehigh University in Bethlehem, Pennsylvania, from the early 1870s until his death in 1895, and his widow gave several gifts to the university in his honor after his death.

From 1860 to 1864 he studied mining engineering abroad at the École des Mines in Paris and the Bergakademie in Freiberg; both institutions included surveying as part of the formal curriculum, though an examination of Coxe's student notebooks from his time at the École des Mines do not give any evidence that Coxe learned surveying while at that institution.[45] Coxe apparently performed much of the surveying of his mines himself, and developed versions of mine surveying tools that were used well into the future. For example, in the early 1870s Coxe designed a plummet lamp for underground work (figure 2.4), that increased accuracy and speed, while requiring only one surveying assistant. The instrument was a small oil lamp in the shape of a plumb bob that hung from a chain. The pointed end of the plummet lamp could mark a surveying point very accurately, and the flame was easy to focus on in a surveyor's transit in order to take surveying measurements.[46] Coxe also invented a steel tape for measuring surveyed distances. Surveyor's chains were cumbersome to use to measure the steeply pitching breasts of Coxe's anthracite mine. He initially conceived of a finely braided steel rope, with small soldered tags indicating distance, but was persuaded by his instrument maker to instead use the raw material for hoop skirt bands—tempered steel, 0.08 inch wide and 0.015 inch thick. A small brass wire was soldered across the tape every ten feet; five hundred feet of the material wound on a reel weighed less than three pounds. Steel tapes of this sort quickly became common in both underground and topographical surveying. Mining engineers who sketched the history of surveying instruments in 1902 noted, "[T]he chain . . . was quite generally used in American mines until Eckley B. Coxe and others started a reformation, some twenty-five years ago, in favor of

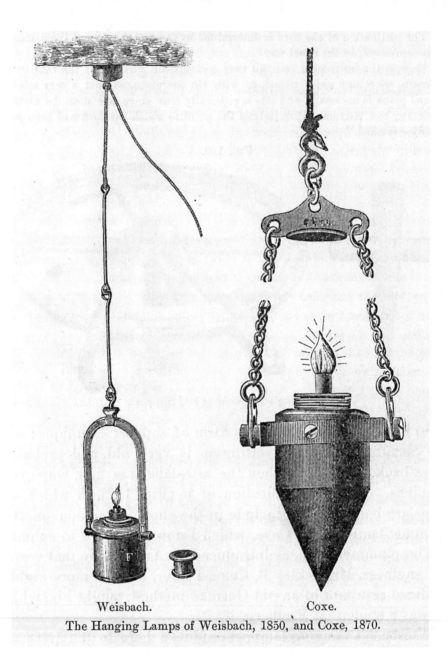

Weisbach.   Coxe.

The Hanging Lamps of Weisbach, 1850, and Coxe, 1870.

Figure 2.4. Eckley Coxe designed the lamp on the right especially for underground surveying, likely inspired by an earlier design created by his mentor Julius Weisbach, shown on
the left. It combined the functions of a lamp and a plumb bob in a single unit, which could
be set up precisely and left unattended, permitting a smaller surveying crew and more-
accurate results. From Dunbar D. Scott and others, *The Evolution of Mine-Surveying Instruments* (New York: American Institute of Mining Engineers, 1902), 304.

the steel band that has now practically consigned the chain to President Cleveland's 'innocuous desuetude.' "[47] Thus, because of Coxe's extensive theoretical training and practical experience in underground surveying, the maps of Coxe's Drifton Colliery examined later in this chapter represent early examples of good underground mapping practice.

Coxe was not an innovator merely in mine surveying, but also in other aspects of anthracite mining practice. By 1878, just two years after Alexander Graham Bell patented the telephone and fascinated crowds at the Philadelphia Centennial Exhibition by showing off the device, Coxe had invested in private telephone lines from his main office to the offices at each of the colliery sites. The state mine inspector noted, "This is certainly a great convenience to all parties concerned," and presumed Coxe's investment in the new equipment would be worthwhile.[48]

Eckley Coxe also took a paternal interest in the health and welfare of his employees that was considered unusually kind by the standards of his time. He built a private hospital for injured mine workers, and created a school for miners and their families. Coxe provided care for employees injured on the job, and provided for widows if miners were killed. He notably installed machine guards on his breaker machinery, at a time when such precautions were unusual and not required by law, to try to lessen the dangers faced by breaker boys. The Pennsylvania state mine inspector noted all these steps with satisfaction, declaring, "[A] better feeling never existed between employers and employes [sic] than does at these collieries."[49]

All of this goodwill did not guarantee labor peace. Along with other coal operators, Coxe endured major strikes of six months or more in each of three years: 1869, 1875, and 1887, as he testified to a congressional committee investigating the coal trade in 1887.[50] The same year, one of his breakers burned down under suspicious circumstances.[51] Strikes and labor troubles, Coxe observed, were "like the measles; they occur periodically."[52] He was a capitalist and had little use for labor unions, preferring instead to negotiate directly with his own workers. Compared to the actions of other coal companies, however, Coxe treated his workers with an unusual degree of respect even when they were on strike. He allowed them to pick usable coal off the refuse piles near the mines, and the miners were able to continue to live in company housing, despite not paying rent. Pointing to one of his striking employees who was in the room during his congressional testimony, Coxe noted, "I have never turned a man out of a house during a strike; Mr. McGarvey here is living in one of them now."[53]

Coxe Brothers & Co. was among the largest independent anthracite mining firms, which meant that it was not owned or controlled by one of the major anthracite railroads, and Eckley Coxe was unafraid of taking on the larger, railroad-controlled companies when he believed it would benefit his firm. Most coal was mined and shipped by companies operated by one of the several railroads that served the region, such as the Philadelphia & Reading, the Lehigh Valley, the Delaware and Hudson, and the Central Railroad of New Jersey, each of which had the economic clout, market power, and control over railroad rights-of-way to dictate frequently discriminatory rates for hauling coal to market. In 1888 Coxe filed a formal complaint against the Lehigh Valley Railroad for excessively unreasonable freight charges. Though it took years for the Interstate Commerce Commission to determine the outcome, Coxe won a rate reduction from the commission in 1891.[54] He bucked the railroads again in 1889. The large companies had formed a combination to restrict output of coal and thus keep prices high, but Coxe refused to play along and sold as much coal as he could mine at lower prices. The railroad companies "all agreed that the Coxe Brothers had been a thorn in the flesh of the combination for a long time."[55] Finally, beginning in 1890, Coxe took a dramatic step toward independence from the railroads by building one of his own, the Delaware, Susquehanna, and Schuylkill Railroad. When completed, his railroad had thirty-two miles of track, connecting all of the Coxe Brothers properties with the tracks of every major common carrier line, enabling Coxe to ship his coal to tidewater or market using his own rolling stock. From mine surveying to railroad construction, market access, and the management of labor, Eckley Coxe was willing to spend money, time, and effort to innovate.

The 1870s were a time of flux in underground mapping practice in the anthracite fields, due to forces including industry consolidation, technological improvement, increasing professionalism of mining engineers, and changing legal requirements. Wide variations existed between mapping practices at individual collieries, but over time mining engineers gradually and unevenly settled on common attributes and features that underground maps should have. Examining a selection of maps from Coxe's anthracite mine at Drifton, Pennsylvania, gives a clear picture of how mine maps could evolve within a single firm.

Eckley Coxe and his brothers started their work at Drifton in February 1865, and were able to ship their first coal in June 1866. A decade later, they brought a second breaker on line at Drifton (which burned in 1887 and was replaced by Coxe's Iron Breaker, seen in figure 2.2). While their mines at Drifton grew, the company expanded to nearby lands in the late 1870s and throughout the 1880s, opening new breakers at Gowen, Deringer, Tomhicken, and Oneida, and substantially refurbishing older operations at Beaver Meadow, Eckley, and Stockton.[56]

The earliest underground map of the Drifton Colliery that has been found, shown in figure 2.5, was drawn in 1870.[57] Coxe had opened his mine at Drifton only a few years before, so this map showed a mine still in its early stages. The same year the map was drawn was the first that coal mines in Luzerne County, including Coxe's, were required by law to be mapped, due to the passage of the statewide mine safety law of 1870. Coxe himself paid close attention to the law, placing it in the context of mining regulations of England and continental Europe for an audience in the American Social Science Association. "I approach [the subject] with some diffidence, as I am personally interested in working mines, and belong to that class from whom the mining law is supposed to protect the miner. I feel, therefore, that my views may not be so unbiased as I should wish."[58] Unlike some operators who chafed at the requirement to map their mines, as we saw above, Coxe had no problem, at least in his public remarks, with the mapping provision of the law. He was concerned, however, that if the mines were to be regulated, that the laws should be enforced by experts who had the education and experience to implement the regulations fairly.[59]

The 1870 Drifton map was drawn in ink on tracing linen, and used colored shading effects to illustrate the breasts; as noted above, breasts were rooms of coal usually worked by one or two miners separated by pillars of coal left standing. The map shows surveyors' stations, unnumbered, connected with thin red lines. Blue arrows, representing airflow, circulate through the works. (Ventilation was a primary concern of the mine safety law.) Doors to regulate airflow are drawn in as well. Air shafts in several places are labeled, and the breasts are long and quite regular in appearance.

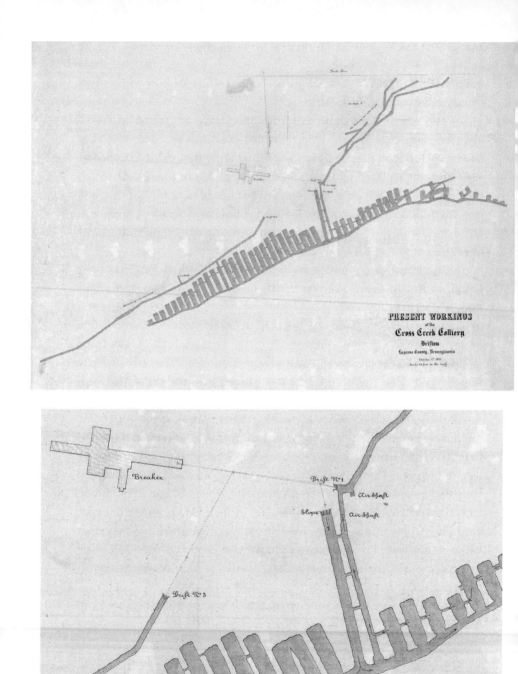

PRESENT WORKINGS
of the
Cross Creek Colliery
Drifton
Luzerne County, Pennsylvania

Surface features, most conspicuously the breaker building, are also visible (see figure 2.6). One section of an unused gangway is labeled "Old Works Now Abandoned"—precisely the sort of dodge that so infuriated Inspector Jones. The magnetic meridian is marked on the map, as is a "Land-Line." This map does not appear to have been altered after it was drawn, probably because of the small size of the sheet.

A map of the same underground space produced two years later shows some different mapping techniques.[60] This map was also drawn in ink on tracing linen with colored shading. A view of the entire map makes it relatively easy to see where this one, from 1872, shows extensions of the workings pictured in the map of 1870 (see figure 2.7). The working edge of the mine is indicated on both ends of the workings, where it says "Face April 24, 1872." No surface features such as the breaker are shown on the map—it represents the underground exclusively. The airflow indicators and air doors of the 1870 map are not represented, although the air shaft is still labeled as such. Perhaps most significantly, surveyor's stations were numbered on this map (see figure 2.8). Numbered stations would ease new surveys of the mine, would make translation of notes from the field into a mapped mine much easier, and would facilitate the transfer of survey information (or job duties) to other people. In other words, this 1872 map is suggestive of a shift from proprietary mine knowledge (since Coxe knew his own mine, he did not need to number the survey stations) to representing the space as an idealized, engineered landscape that other mining engineers could understand.

As explained above, the mine safety law gained harsh (though rarely enforced) penalties for poor or nonexistent maps in 1876, and mine inspectors strove diligently to share information about good mapping

FACING PAGE:

Top: Figure 2.5. This map of Eckley Coxe's colliery at Drifton was made in 1870, the same year Pennsylvania required anthracite mines to be mapped. "Present Workings of the Cross Creek Colliery, Drifton, Luzerne County, Pennsylvania, October 1st. 1870." Map 14-22, Collection 1002 Coxe Brothers Collection, Archives Center, National Museum of American History, Smithsonian Institution.

Bottom: Figure 2.6. Detailed view of a portion of the map from figure 2.5, showing the breaker building, shaded drawing effects, and unnumbered surveyors' stations. Map 14-22, Collection 1002 Coxe Brothers Collection, Archives Center, National Museum of American History, Smithsonian Institution.

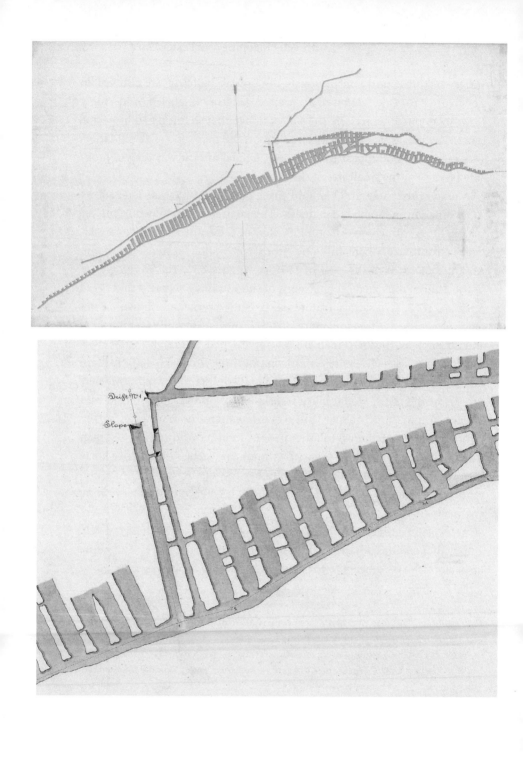

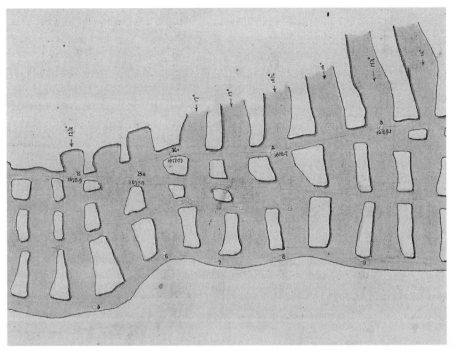

Figure 2.9. Detailed view of a portion of a map from 1879, showing multiple numbering systems for surveyor's stations, elevation measurements, and dip (steepness) measurements of breasts. "Map of a Portion of the Cross Creek Colliery, Luzerne County Pa, 1879," Map 14-27, Collection 1002 Coxe Brothers Collection, Archives Center, National Museum of American History, Smithsonian Institution.

FACING PAGE:

Top: Figure 2.7. Map of Eckley Coxe's mine at Drifton two years later. Note the visible extension of the mine, compared with the 1870 map of figure 2.5. "Tracing of the C.C.C. Workings Slope No. 1 showing extension of working [*sic*] April 1872," Map 14-23, Collection 1002 Coxe Brothers Collection, Archives Center, National Museum of American History, Smithsonian Institution.

Bottom: Figure 2.8. Detailed view of a portion of the map from figure 2.7, showing numbered surveyor's stations. Compare with figure 2.6. Map 14-23, Collection 1002 Coxe Brothers Collection, Archives Center, National Museum of American History, Smithsonian Institution.

practices. A detailed look at a map made in 1879, shown in figure 2.9, shows more information being added to the maps.[61] This map was also drawn with shaded relief. Surveying stations are shown, but the mixed use of letters and numbers suggests overlapping surveys. This map differs from the 1870 and 1872 maps in that it shows some elevation measurements, usually noted alongside surveying stations. The elevation measurements are four-digit numbers, often with a single place past the decimal as well, and indicate elevation above average sea level.[62] Many Pennsylvania anthracite veins pitched quite steeply, and elevation notations (or some other way of indicating vertical relief) permitted a much more accurate representation of the physical geography of the underground.

Elevation measurements on a common datum were long a goal of professionally minded mining engineers in large districts, to help avoid serious problems when two mines approached a common property border. Anthracite mining engineers were early adopters of elevation above mean tide measurements, but this was greatly facilitated by the fact that the anthracite mines were closely serviced, and often owned, by railroads and canal companies who had tidewater operations. Railroads and canals were among the earliest users of carefully measured elevation data, to ensure that track grades were not too steep for the small locomotives of the time to handle and that water would keep canals full. Since railroads and canals kept careful measurements of elevations above tidewater for points along their lengths, it was easier for anthracite engineers to tie their surveys into these points of known elevation. To facilitate this process, the *Anthracite Inspectors' Reports* published tables of known elevations provided by railroad and canal companies in 1870, and again by request in 1871.[63] By contrast, many mines in remote regions used the collar of their main shaft, where earth met air, as the zero elevation for their maps. This would be convenient for each individual mine, but made comparing surveys of different mines a tremendous burden because of the need to recompute all of the elevations. Thus, by 1879 the information inscribed on underground maps had become more extensive, and had begun to include potentially useful information about the underground environment.

Maps of underground landscapes often showed only the single property under consideration, but could also be used, like surface maps, for indicating property lines and facilitating action on boundary disputes. A series of maps made by the Coxe operation between the 1880s and 1905 were intended to articulate the company's side in an ongoing boundary dispute with the Highland Colliery, which operated on an adjacent tract of land.[64] The map from 1889, a portion of which is shown in figure 2.10, appears similar to the 1879 map in figure 2.9, with shaded effects, numbered surveyors' stations, and elevation measurements. This map also featured, in a pink tint, a proposed boundary between the properties. This boundary pillar, made up of coal, was intended to provide a physical barrier to prevent water from the Highland workings from flooding the lower Coxe works. However, because of excavations that had already taken place, it was necessary to make the physical boundary deviate from the actual property line. Coxe proposed a jagged boundary pillar whose zigzag course resulted in identical amounts of coal credited to each company, and an even split in ownership of the useless barrier coal. The map was used as a tool to calculate the area of coal transferred, retained, and lost, as can be seen in figure 2.10.

The importance of the map for this calculation is clear. A more accurate estimate would have measured *volume,* rather than area, but the map as drawn in two dimensions did not make volume calculations easy. The amounts of the coal (in square feet) were inscribed on the relevant portions of the map, and a narrative comparing the two sides was added near the map legend. These efforts to persuade Highland to accept the plan of the Coxe Brothers apparently paid off; later maps show the pillar as the agreed-upon line. Other maps in the series, such as one from the 1890s shown in figure 2.11, were used to monitor encroachment on the pillar. The workings of the Highland Colliery are seen working into the parallel lines of the boundary pillar. Highland's actions would have weakened the safety pillar and increased the risk that water from the higher Highland works might burst through into Coxe's mine.

After the turn of the twentieth century, surviving maps of the Drifton Colliery take on a more dynamic quality, as mapping became an expected process throughout the mining industry and maps became more integrated into the daily work of the mine. The period was a time of turmoil for the company, with the death of Eckley Coxe in 1895, major strikes in the years around the turn of the century, and the sale of the Coxe Brothers & Co. operation to the Lehigh Valley Railroad in 1905.[65] The company continued to mine and market coal, but some of the management changed, and may have brought different mapping techniques with them.

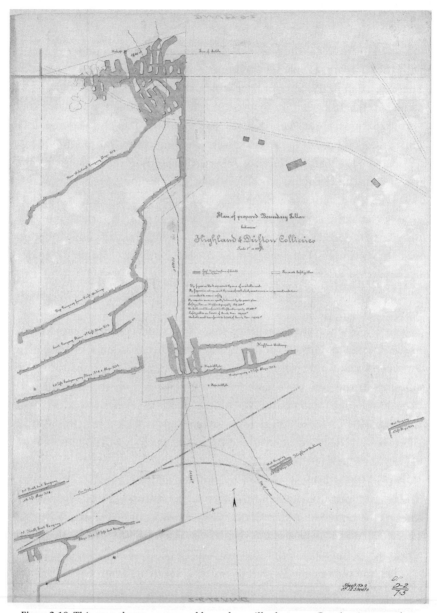

Figure 2.10. This map shows a proposed boundary pillar between Coxe's mines, on the left side, and the adjacent Highland Colliery. Because both companies had already mined close to the boundary, it was necessary to draw up a zigzag pillar to ensure that the loss of mineable coal, locked away in the safety pillar, impacted both companies equally. Calculations of how much coal would be lost by each were inscribed directly on the map. "Plan of Proposed Boundary Pillar between Highland & Drifton Collieries" (Sheet 2 of 12), Map 14-16, Collection 1002 Coxe Brothers Collection, Archives Center, National Museum of American History, Smithsonian Institution.

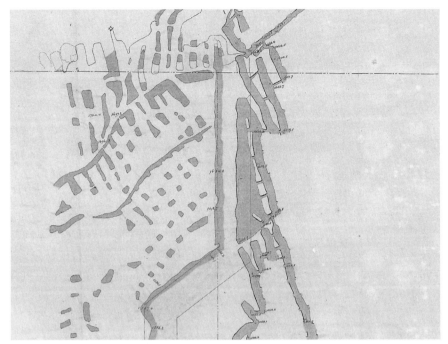

Figure 2.11. Detail from a later map shows the Highland Colliery's works extending into the agreed-upon safety pillar, making it thinner and weaker, increasing the possibility it might fail and allow water from the Highland to flood the lower Coxe works. "Sheet 4 of 12," Map 14-16, Collection 1002 Coxe Brothers Collection, Archives Center, National Museum of American History, Smithsonian Institution.

Many twentieth-century maps show revisions of a decade or more on a single sheet. A map of the Buck Mountain vein of Drifton No. 2 shows evidence of being in use as of 1905, with periodic revisions between 1914 and 1927 (figure 2.12).[66] This map, six and a half feet long, clearly shows the evolution of such documents to store greater amounts of qualitative and quantitative information.

This map is a simple line drawing, lacking the shaded perspective of earlier maps, and was executed mostly in black ink, but has at least one notation in color, as well as several in pencil plus erasures. The lack of shaded features would have made such a large map faster to draw and easier to reproduce with traced copies or blueprints.

The company inscribed a considerable amount of geological information useful to their operations on this map. For example, the map contains diamond drill hole notations. Some are evidently from an earlier round of

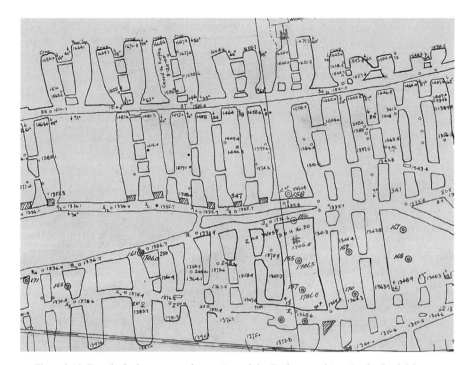

Figure 2.12. Detail of a large map of a portion of the Drifton workings in the Buck Mountain Vein, which was used over more than two decades from circa 1905 to 1927. The simpler style of drawing would have made this map easier to blueprint. Compared with earlier maps, far more information is recorded here. The unstable working conditions and steeply pitching veins can be recognized in the terse statements of the engineers on the map. "Drifton No. 2. Buck Mountain Vein, West End Workings, Bottom Bench," Map 14-10, Collection 1002 Coxe Brothers Collection, Archives Center, National Museum of American History, Smithsonian Institution.

diamond drilling and occasionally feature notations on the depth to coal and the size of the coal seam, but others are marked only by a number and a bulls-eye feature. Cropping and pinching of the coal vein was noted directly on the map. Many points on the map, especially in the breasts, show measurements of degrees with an arrow. These indicate the dip of the vein, and are a suggestion of how steep the works are in that area. In some breasts two arrow/degree measurements were used, which indicate both dip and strike.

This map gives some good examples of the steeply pitched breasts needed to extract coal from this mine. For example, one breast near the center of the portion of the map shown in figure 2.12 has an elevation at the top of 1,464 feet and a pitch of fifty-eight degrees downhill. If the dip was constant all the way to the bottom (a tricky assumption, given the twisted folding nature of the vein), the breast would be approximately 250 feet from the feed door at the bottom to the very top. Some breasts on the map such as this one appear to have an obstruction in them, marked by cross-hatching. These are likely working breasts with the feed doors at the bottom still in place. In breasts that rose at a steep enough angle, miners did not have to shovel coal into cars. Instead, they built a loading chute at the bottom of the breast, and used timbers to create a tiny hallway to one side of the breast, called a manway, for the miners to get in and out of the working space.[67] In these types of spaces, the miner would simply blast down the coal, and it would fill the loading pocket formed by the doors at the bottom of the breast. The next day, the miner would stand on the broken coal to reach the top of the breast to blast down fresh coal. Since broken coal took up more space than solid coal, the miner would occasionally draw off coal from the bottom to ensure he had a platform of optimal height to work the unmined coal above. Once the breast had been fully excavated, the miner would withdraw the remaining loose coal through the feed doors in the bottom of the breast, directly into coal cars. The time to work a breast to completion could vary from several months to a year, depending on the geology and the size of the chamber.[68] In figure 2.12, the surveyor's stations are usually along an edge of the breast, which indicates the presence of a manway. Also note the consistent elevation of the gangway at the bottom of the breasts, which would facilitate transportation of mined coal in cars pulled by mules or small locomotives.

This map also gives a considerable amount of information about the work environment of the mine, particularly along the edges of the works as the coal seam approaches the surface and work becomes more hazardous. By this time, the company was actively engaged in robbing the pillars in older workings, as indicated by notes such as "To West Line of George Moore [Tract] Top and Bottom Robbed Oct. 1905."

Many breasts on the map, such as those at the top of figure 2.12, terminate with the term *crop*, but one has the notation "Caved to Surface, 3-20-05," and another says "Poor Top." The unstable nature of the mine is evident in many places elsewhere on the map. A surface cave-in is indicated by a circular feature with hatchures, labeled "7-12-14, 32' Deep,"

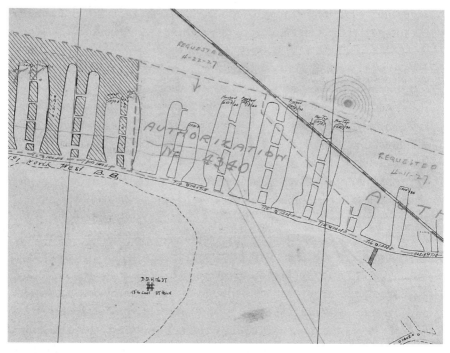

Figure 2.13. On the same map as figure 2.12, engineers sketched out plans to rob the pillars, which involved removing the coal from the supporting pillars and letting the roof collapse. Such an operation was dangerous and could damage the structure of the mine. Each authorization was requested via a series of paper files, which are linked to the map through reference numbers. Map 14-10, Collection 1002 Coxe Brothers Collection, Archives Center, National Museum of American History, Smithsonian Institution.

and a different portion of the map contains some indeterminate-looking surveys and the note "7-2-14 Not Safe to Go Farther." Many notations on individual breasts—"Roll," "Full Vein Roll," "Pinched," "Pinched Full Vein," "Fault," and "Fault Full Vein"—explain the geology of the coal seam. Taken together, these notations suggest the power of the map to collect and store important information about the always-hazardous underground environment.

Larger robbing projects were indicated, as in figure 2.13, by a dotted line in orange ink surrounding an area on the map, with "Authorization" and a number. A future authorization appeared on the map first in pencil, with the notation "Requested 4-22-27." This authorization number referred to a series of forms and blueprinted partial maps, stored in the

files of the engineering office, that detailed the probable damage and production yields of the proposed robbing operation. Once approved by management, miners were permitted to commence the robbing.[69]

Looking at a span of maps from 1870 to 1927 of a single mine in a
single place can clarify some long-term trends relating to the visual culture
of mining. Most importantly, the maps became useful as a representation
of the underground. The early maps would have been of little use to someone not already intimately familiar with the Drifton works, but the later
maps provided enough information to enable someone unfamiliar with
the place but familiar with the engineering language they employed to
have a good appreciation of the status of the mine. As a result, the most
powerful knowledge associated with the first maps—local knowledge, of
the sort possessed by miners—gave way in importance to engineering and
management knowledge, which harvested data about the status of the
underground; encoded it in numbers, symbols, and words; and inscribed it
on a map. This shift is associated with the increasing utility of the map as a
basis for decision-making by managers who were removed from the immediate context of underground work.[70] It is also evidence of and a tool for
the solidification of engineers' control over underground work. A skilled
miner in 1870 was more likely to be able to make the leap into management's ranks than was his counterpart in 1927; the gap in the latter case
consisted in part of a professional education, and a specialized, visual, language of representation and control.[71]

The Coxe maps also illustrated the increasing capability and use of
visual representations to store a wide variety of data, particularly quantitative data. These data took two forms: The first form was hypertextual. Over
time, the maps increasingly referred to other documents—first, surveyor's
notebooks, but later work authorizations, borehole logs, and the like. The
other trend was the increasing amount of data stored on the map itself.
Later maps were not merely two-dimensional pictures of a bird's-eye view
of the underground (as was the 1870 map in figure 2.5), but also described
geology, elevation, and work controls, and provided fine-grained information about the stability of the workplace.

The prominence of the Pennsylvania anthracite region in American mine mapping practice lasted for no more than a couple of decades,
but in that time mapping became a central part of the duties of mining engineers. By the turn of the twentieth century, young graduates of

American mining schools learned surveying as part of their educational requirements, and practiced it in whatever mines they found themselves working—not just in the anthracite mines, but in the bituminous coal mines of Appalachia, the copper mines of Michigan, the gold and silver mines of the West, or in places as far-flung as Mexico, South Africa, and Asia.[72] In every case they brought the visual culture of mining with them— a set of artifacts, discourses, and practices that helped them create and control underground geographies of mineral production that provided the raw materials for the Industrial Revolution. Chapter 3 will examine that visual culture put to work to describe, understand, and defend the copper mines beneath Butte, Montana.

chapter 3

New Maps, the Butte System, and Geologists Ascendant

Butte, Montana, has been called the "richest hill on earth" for the remarkable concentration of minerals in its rocks. These minerals underpinned a mining town of remarkable longevity, culture, and labor activism. Though the mines have largely disappeared and a massive open pit with a toxic lake in the center has swallowed many of its neighborhoods, Butte still shows off a rugged charm and a keen sense of its own history. Much of the town has been declared a national historic landmark (which is remarkable in that it is, in fact, toxic) to honor the role played by the site in the extraction of the raw materials that helped support America's twentieth-century technological ascendance.

Butte is a historic site in another way—it was there that a new way of mapping and thus understanding underground mines developed and spread throughout the mining world. This new system used the visual culture of mining in a new and powerful way. The so-called Butte System or Anaconda System (the latter name after the company who created it) was an evolutionary leap forward in mine mapping in several senses. First, the Butte System resulted from a reconceptualization of the physical form of a mine map, breaking it apart, blowing it up, reproducing it, and even rethinking how it would be stored, in order to integrate it more fully into a revamped system of underground knowledge. Second, the Butte System represented the widespread and systematic collection and storage of geologic data on underground mine maps. While earlier mine maps would sometimes note features of geological interest, the Butte System maps made the display of geology as important as the display of the workings. Finally, the Butte System of mapping facilitated an important reorganization of authority over mining operations, placing geological knowledge, embedded in an engineering context, at the center of decision-making. Companies had occasionally consulted geologists before, of course, but in Butte the work of the mining geologists—usually trained engineers with additional geological expertise relevant specifically to mining operations—became central to the operations of Anaconda, with implications for the work of mining engineers and others.

The development of the Butte System was, especially in the early years, a response to a host of conditions—some common to mining in the American West, and some peculiar to Butte itself. Butte's complex veins and legal environment ripe for trouble provided the preconditions for innovation in underground mapping.

Butte began as a gold placer mining camp in the 1860s. An initial rush for placer gold gradually gave way to lode mining, mainly for silver, in the 1870s and 1880s. Though silver and gold were what most miners were after, the ores were also very rich in copper, and by the mid-1880s miners in Butte were mining copper from underground veins. The ores were also very rich in other minerals such as manganese and zinc, which became economically important in later decades. The Anaconda Mine, which had been initially exploited for its silver, discovered large reserves of extremely rich copper ore (chalcocite) at greater depth in 1882, and under the leadership of Marcus Daly, systematically created a comprehensive copper mining operation.[1] Many other companies in Butte followed suit.

As was the case in many Western districts, mining claims in Butte were governed by the federal Mining Law of 1872.[2] Among other provisions, this law determined the size of claims; and the requirements for staking a claim, holding possession of it, and converting it to private property. It also contained a clause—the source of extraordinary legal trouble across the West—that permitted the owner of a claim to follow a vein even if it veered underground into neighboring property, as long as the apex of the vein was on the original claim. This so-called "apex law" (which we will also cover in detail in chapter 5) provided a ripe environment for lawsuits in any Western mining district.

The particular geology of Butte, however, made the district extremely susceptible to apex lawsuits. Extensive faulting had shattered and moved all of the major Butte veins, and multiple waves of ore deposition had created tiny networks of vein material in between the primary veins, which were sometimes faulted again.[3] Furthermore, the rock cut by the veins was all similar-looking granite, so it was even more difficult than usual to find veins that had been shifted out of sight.[4] Thus, tracing a vein in a continuous fashion underground from its surface apex was especially challenging in Butte, but this sort of continuity was vitally important to sort out the ownership of the rich veins because of the terms of the apex law.

The requirements of the apex law eventually gave geology a greater importance in Butte than was usually the case elsewhere. Creating an underground mine was a feat of engineering. Engineers (and perhaps even miners) needed only a fairly casual, practical understanding of geology—enough to distinguish good ore from bad, and to process it in the mill. Following veins, especially after faulting, was an art, but again relied little on geological theory and more on a practical, experiential geological understanding of the underground. Deeper geological questions, such as those about the origins of rocks, were rarely of much consequence to nineteenth-century miners and mining engineers. If the ore was within sight in the mine, they needed to extract it; if it was not in front of them, they needed to find it; but geological theory simply had little relevance to these operations.[5] As one mining observer opined in the 1860s, "Science and experience do not appear to give much assistance in prospecting for quartz lodes. Chemists, geologists, mineralogists and old miners, have not done better than ignorant men and new-comers. Most of the best veins have been discovered by poor and ignorant men. Not one has been found by a man of high education as a miner or geologist. No doubt geological knowledge is valuable to a miner, and it should assist him in prospecting; but it has never yet enabled any body to find a valuable claim."[6]

This might have been true generally, but geological knowledge was quite valuable in one specific situation. The Mining Law of 1872 declared that a miner could follow all "veins, lodes, and ledges" whose apex was found on the miner's claim. A vein is a geological reality, recognizable in the ground in an ideal case (such as that imagined by Congress when the law was passed), but Butte Hill and many other locations throughout the West posed nonideal cases.

This is where geological theory became important to practical miners and engineers. Tales of secondary enrichment or meteoric waters might not help the miner underground find rock with metal in it, but it could make a world of difference above ground, in the court house.

Dating the advent of the everyday use of geology by the mining industry is challenging. It is clear that geology suffuses into mining engineering over time, but marking the specific moments when that happens can be very difficult. There are occasional moments, however, where it is possible to see a clear relationship between those who know the geology and those who direct the underground work.[7] One set of those relations occurred in Leadville, where S. F. Emmons, a mining-oriented scientifically trained geologist working for the US Geological Survey (USGS), produced a geological report that was of immediate and practical use to help find ore.[8] Emmons' report helped make Leadville a camp that amply demonstrated the value of geological knowledge.

One person in Leadville who must have been watching was a young mining engineer named David W. Brunton. Brunton was a Canadian-born inventive genius and engineer, trained at the University of Michigan, who arrived in Leadville in 1879 at age thirty.[9] During his tenure in Leadville, Brunton served as a consulting engineer for mines involved in apex litigation. (Brunton also perfected the art of mechanically sampling ore for assaying, an innovation that first made his name and reputation in the mining industry.)[10] Owners of several mines in Aspen, Colorado, lured him away in 1886 because of his expertise in carefully preparing mines to present facts for litigation. Brunton first came to Butte in 1890 on a temporary consulting basis, to direct the underground development work for the Little Darling Mine in its litigation with the Blue Bird.[11] In Butte Brunton caught the eye of Marcus Daly, owner of the Anaconda and other mines. In late 1897 Daly asked Brunton to come to work for him, as Daly's Butte mines had serious litigation problems of their own. Brunton remained a consulting engineer for Anaconda and its successor, the Amalgamated Copper Company, until 1905.[12]

Daly's mines were a challenge to engineer. They were fantastically rich and were being worked at a breakneck pace. Historian Michael P. Malone described the Anaconda of this era as a "leviathan" of a mine, a "giant, highly integrated organization."[13] Since Daly's mines were under threat of litigation, Brunton had to develop the mine in such a fashion as to pay the bills but also to provide geological evidence that would help their side in court. To add to the complexity of Brunton's work as a consulting engineer, Daly kept acquiring new properties, adding them to his stable of mines.[14] Thankfully, Daly gave Brunton clearance to add to his staff to cope with the workload of preparing for litigation.

Here Brunton made an important choice, one that was rooted in the specific circumstances of laws and faults, but that ultimately had far-reaching ramifications for the practice of underground mapping. He hired Horace V. Winchell, a geologist, to help him work on Daly's Butte mines and prepare to defend them in court. Clearly, Brunton had an inkling of how important geological information could be, especially in a pending lawsuit. Perhaps his formative experience in Emmons' Leadville and in the Aspen lawsuits shaped his thinking at this critical point. Whatever the reason, Brunton seemed to have a greater appreciation of geological information than did most of his contemporaries. Decades later, Brunton was remembered as "one of the first of the mine operators in Butte to realize the advantages to be gained in the use of mine geologic maps and models as practical aids to daily mine operations."[15]

Brunton chose well. Horace V. Winchell was born in 1865 into a family of geologists. His father and uncle were both university professors and state geologists, of Minnesota and Michigan respectively. Winchell's geological interests ended up on the practical rather than theoretical side of the geological spectrum. Even before completing his university degree at the University of Michigan, he received an early apprenticeship on the Minnesota State Geological Survey, concentrating on the iron ore ranges; in a series of coauthored and single-authored reports in the early 1890s, he described the Mesabi iron range and made bold predictions of its future productivity, which eventually proved true. Attempting to parlay this success, Winchell opened a consulting geology business, headquartered in Minneapolis, but had only a handful of clients and contacts.[16]

Winchell met Brunton, then a vice president of the AIME, during the July 1897 AIME meeting in the Lake Superior region. As part of the local arrangements committee, Winchell guided the mining engineers around the iron ranges of northern Minnesota.[17] Early in 1898 Winchell met Brunton again, and the latter attempted to entice Winchell to come to Butte. As Winchell related the story, he was reluctant to do so—"I promptly declined, saying that I had assured my wife that there was one place in the world where she would never have to live, namely, Butte," but Brunton persisted and Winchell eventually caved.[18]

The beginnings for Winchell in Butte were modest. Winchell apparently intended to spend only a year there—just another consulting job—not realizing that Butte would become his home for eight years. Similarly, Winchell later noted that there were no plans at first—by either Winchell or Anaconda—to develop the sort of geological enterprise that eventually evolved there. Instead, beginning in 1898 Winchell spent two years learning about Butte mines and helping to "look after certain developments" in a part of Daly's Anaconda that was entangled in litigation with a neighboring mine.[19]

Winchell organized the geological department at the company in 1900.[20] The department at first extended Winchell's original efforts to understand the complicated geology of the Butte District, to direct underground construction intended to systematically expose geological evidence important in lawsuits, and, perhaps most importantly, to comprehensively and systematically record geological data found underground.

A tidal wave of lawsuits over mine ownership swept over Butte from 1898–1906; observers and historians termed these lawsuits as "war" or "battle," fueled by outsize personalities struggling for total control of Butte at any cost. Described by historian Michael Malone as "an eye-gouging, knee-to-the-groin fight," the political corruption, nefarious activities, and callous manipulation of the citizens of Montana has been well chronicled.[21] One of the forces animating this struggle was the attempt to consolidate Butte's copper mines under single ownership. In 1899 Daly's Anaconda became the centerpiece of the Amalgamated Copper Company, a holding company organized by several Eastern capitalists with close ties to Standard Oil, with the intent of eventually creating a copper trust to dominate the industry, set prices, and reap enormous profits. (Anaconda continued to exist as an operating company within the Amalgamated umbrella, as did other mining companies Amalgamated controlled, until all of Amalgamated's Butte holdings were transferred to Anaconda in 1910. Other Amalgamated holdings across the US were shifted to Anaconda in 1915 and Amalgamated dissolved, leaving Anaconda the owner and operator of the huge enterprise Amalgamated had assembled.)[22] Daly himself, and many of his close associates, remained with the new company at first, though Daly sold his shares in 1899; he died shortly thereafter, in late 1900. Within a year or two, Amalgamated acquired other Butte copper mines, such as the Boston & Montana group and the Butte & Boston mines. Rival Butte capitalist F. Augustus Heinze, operating with "unscrupulous ingenuity" (to use Malone's evocative phrase), worked tirelessly to fight off Amalgamated, tying up the company's mines in lawsuits, buying off judges to secure favorable court decisions, and even resorting to secret mining of disputed ore underground in flagrant violation of court injunctions.[23] (When Winchell and his assistant geologist Reno H. Sales discovered those violations in late 1903, the situation devolved into open conflict in the mines, complete with fistfights between rival crews, improvised weapons, booby traps, dynamiting, and homemade smoke bombs.)[24] As the company made the standard mining lawsuit preparations of hiring outside experts and lawyers seemingly by the dozen, Brunton and Winchell girded for combat by deploying the visual culture of mining—surveying, mapping, recording, plotting, and modeling Butte's underground at an almost unprecedented level of detail.

Brunton, working with Winchell, brought an innovative system of geological mapping to Anaconda just before the turn of the century. Brunton had developed the basics earlier in his career, but his system gained lasting fame only as part of the Anaconda operations in Butte. Brunton and Winchell's new system of mapping had its roots in new techniques Brunton had tried in Aspen, to map faults in the mines and help miners find veins after they had been faulted out of sight, but the Brunton/Winchell partnership at Butte advanced techniques rapidly.[25]

Brunton saw the traditional composite mine map with fresh eyes, in a way that allowed him to extend its usefulness. The first step was to reimagine the underground mine survey. Brunton noted that the eventual use of mine maps was not commensurate to the amount of money and effort expended in making them. Engineers might occasionally check them to determine the relative position of mine workings or boundary lines, but made little more use of them. As a proponent of forward-thinking progress driven by science and technology, Brunton verbally swiped these maps into the proverbial dustbin: "The ordinary mine-survey map, being nothing but a record of what has been done, is, in one sense, only ancient history."[26]

Brunton instead recommended creating a new system of information about the underground, by adding geological information to the existing mine maps in a way that it could be used in making decisions in both operations and development. Brunton's innovation, that Winchell and Reno Sales at Anaconda later developed further, was to take ordinary large-scale mine maps, trace them on to semitransparent vellum, then have geologists plot fine-grained geological information on the maps based on the rocks exposed in the mine. The individual maps were bound together in a loose-leaf system, making folios that covered small parts of the mine with each sheet marking increased depth. The exact relation of the sheets to one another, combined with their semitransparency, enabled users to see the geology of each level through two or three sheets, giving a poor (but inexpensive) imitation of the sort of representation created by glass models.

To create these new visual representations and make them useful, the existing maps had to be reworked into a new information system, complete with different workers to collect and record different sorts of underground data. The first step, according to Brunton, was to disassemble the map. From the composite map, which showed all of the workings of the mine superimposed on one another in a tangle of lines, Brunton traced sheets for each individual horizontal level. He noted that the forty feet to one inch scale, common on maps of metal mines at the time, was convenient. Brunton took extreme care when tracing each level sheet, so that the finished sheets lined up perfectly atop each other. Each sheet was then hole-punched, and the maps were placed together in a wooden holder like a type of oversized three-ring binder that held them in proper vertical position so that the workings would line up from one sheet to the next. Brunton was careful to specify that the maps should be made on tracing cloth. This was not only to facilitate their initial creation from the general mine map, but also because when bound together in the folios it was possible to see several levels at once through the semitransparent material, making it easier to project new workings and construct vertical sections. These folios of horizontal maps were to be kept in the office, and Brunton noted how conveniently each set would slide on its wooden backing into a cabinet with horizontal runners, like a chest of drawers with the drawers removed.[27]

After Brunton disassembled the general map into horizontal level maps, they made destructible copies to take into the mines. At first, he tried transferring them into notebook-sized sheets by tracing over the level maps with carbon paper underneath. These notebooks—Brunton found that larger notebooks sized eight by ten inches were particularly convenient—further broke the level maps apart into manageable sizes. However, their primary purpose was to serve as a canvas for geological notation in real time. Brunton later tried lightly printed blueprints and brownprints, and found that, especially "where the geology is reasonably simple," these copies, folded up and stuffed into a convenient pocket, worked just as well under most circumstances. Brunton's caveat about complicated geology probably stemmed from the inability of blueprints to reproduce color, which was ordinarily how geological information was portrayed on the level maps. In this case the inexpensive and destructible nature of blueprints made them perfect for carrying "directly into the mine."[28]

Down in the mine, geologists would note as much information as possible on the sheets, particularly faults, orebodies, and related information. "Nothing that can be seen by the naked eye is too small or of too little importance to be noted and recorded," related Winchell.[29] Resurveying old workings for the geology was a tedious task that had to be done carefully. Shifting development in the mine might render workings inaccessible over time, so the work had to be done right. Additionally, in preparation for lawsuits the mining geologists might have an opportunity to visit rival mines owned by a different company, to map and sample them as well, but there was no reason to expect such access once the trial at hand was concluded. Keeping up with new workings as they advanced, however, was much easier. In part this was because it was easier to see geological phenomena in freshly exposed rock, but Brunton also discovered that it was possible to enlist the help of the underground foremen and miners in the project. "In fact, it will usually be found that the foremen, shift-bosses, and even the common miners will take such an interest in the work that every change of rock or ore-values, and every slip or fault will be pointed out to the geologist by the men, instead of his having to hunt for them, as he has had to do in the older workings."[30]

Data on underground geology from the mine sheets and notes would be inscribed on the folio maps to bring them up to date. Additionally, Brunton recommended recording information from underground assay samples, though he cautioned against burdening the map with too much information. Instead, only a number should appear on the map, which in turn referred the user to a notebook where more-detailed records of each sample were kept.[31] From these data-rich horizontal folio maps, Brunton constructed precise vertical geological sections. These sections more vividly portrayed the amount and quality of the ore left in the ground, and made the development of the mine more immediately visible than was the case on the horizontal maps. The construction of vertical sections also provided a visual check that the geological data recorded on the horizontal maps were accurate, because errors would look out of place at first sight.[32]

To conduct this new work, Brunton encouraged big mining companies to hire a new type of professional, the mining geologist. The mining geologist, as he described it, would "continuously" keep the new maps up to date with progress underground. Brunton cautioned that the work he described should be the only responsibility of the mining geologist, so he could pursue it "with industry, enthusiasm, and technical knowledge." Thanks to geological training and having no other responsibilities, Brunton's mining geologist could find and record "with precision those indications which hard-worked superintendents, foremen and surveyors, however intelligent, might easily overlook," and would do that work with "engineering accuracy" in order to maximize its usefulness.[33]

This extensive data recording and mapping effort was justified initially because of the pressure faced by the companies from lawsuits. The principle of extralateral rights, embedded in US mining law, held that if a mine controlled the apex of a vein, it was permitted to follow that vein downward into a neighboring mine's property. (Chapter 5 discusses extralateral rights at length.) The upshot was that in an extralateral rights case, property title might hinge on a geological question, and in Butte the underground geology was very, very complex. All of the careful measuring, recording, and mapping were critical to the Anaconda's legal defense. "In this way the mines in controversy are studie[d] until the structure is thoroughly understood and the theory of defense or offense formulated."[34] Once a mine had been rendered thoroughly visualized and geologically coherent by the maps of the geologists and engineers, the geologists closely directed additional excavation work to provide further evidence to strengthen their theoretical stance. "Work plans and sections are prepared for the use of the mine foremen showing exactly what openings are to be made and orders are given that the work be rushed by two or three or four shifts of men, under the personal inspection of the geologists," recalled

Winchell.[35] All of the results of the underground work would be made a part of the visual information record of the mining operation, and from those records elaborate maps and models were made for courtroom use, to reinforce and highlight the geological concepts that would determine the ownership of the mine.

The period where lawsuits were the major focus of the geological department at Amalgamated lasted into 1906. In February of that year, Butte capitalist Heinze sold his interests to an Amalgamated-controlled holding company, and, importantly, settled the 110 lawsuits against Amalgamated his companies had filed.[36] Heinze's sale also served to consolidate most of the productive part of Butte under Amalgamated's control, opening the way for efforts to work all of the Butte mines in a systematic, integrated fashion, guided by the Amalgamated/Anaconda geologists. Winchell departed the Amalgamated shortly thereafter, taking a position as chief geologist for the Great Northern Railway.[37] Reno H. Sales, Winchell's bright assistant geologist who he had hired in 1901, was promoted to replace him, and Sales spent the rest of his long and distinguished career perfecting the Anaconda System of mapping and applying geological understanding to mining problems.

Having paid a high cost to achieve a greater understanding of the geological origins of Butte's riches, the mining geologists turned that knowledge directly to the business of finding more ore to mine. A longstanding article of faith among mining geologists is that the geological work for the lawsuits directly resulted in new discoveries. Speaking in 1906 to a graduating class of mining engineering students, Winchell said so explicitly: "Law suit developments resulted in the discovery of many valuable ore deposits in Butte, and may perhaps have paid their expenses in so doing."[38] Winchell's successor Reno Sales, concluded differently—"I think I can safely say that the actual development work done for litigation purposes didn't find any more ore," he told an interviewer in 1961.[39] Sales pointed out that the Butte lawsuits were always about orebodies that had already been found. However, the extraordinary preparation and precision that was needed to prepare geological information for lawsuits was "a very fine background for geological work" that helped discover new ore in Butte throughout the twentieth century.[40]

The precise geological work of the Anaconda Company rested on extensive, accurate, and rapid surveys of the mines by Anaconda mining engineers, to create the necessary foundational maps of the works. In 1910 Paul A. Gow, a mining engineer at Butte, described the methods he used in surveying the mines.[41] His surveys tied underground and surface together in a coordinate system.[42] Each new mine survey began from a carefully constructed baseline in Butte, some five hundred feet long, over ground that had not been undermined. The length and undermining precaution were intended to ensure that the line would remain a reliable length and its elevation would not change. This baseline was marked with iron poles set in concrete. The elevation above sea level of the line had been carefully determined, which permitted Gow to calculate underground elevations with a little simple math.[43]

Bringing these known coordinates underground was no easy task. Ideally, Gow plumbed two different shafts that both connected to the underground level to be surveyed. The separation between the shafts, likely several thousand feet or more, ensured that any small errors of measurement could be worked out with ease, since the base of the survey was so large. If a second connecting shaft was unavailable, Gow used two plumb wires (one at a time, to avoid tangling them), separated as far as was possible within the single shaft itself. Because in this latter case the base of the underground survey was so small, only three or four feet, Gow noted the substantial precautions, including repeated measurements, that surveyors must take to minimize possible errors in measurement.[44]

Interestingly, in the levels of a metal mine, Gow noted that he measured only horizontal distances, taking note of vertical angles only in the special case of mapping an interior vertical raise or connection. Presumably it was reasonable to not record vertical angles because the levels themselves had almost no vertical relief, and the ability to ignore a set of additional measurements doubtless speeded up Gow's work. (By contrast, no such assumption could be made in a coal mine.) When it was necessary to survey a raise, he used the top or side telescope of the transit, as the ordinary telescope was blocked in measuring an extreme angle. (The Heller & Brightly mining transit pictured in chapter 1, figure 1.1, features a top auxiliary telescope and would have resembled Gow's instrument.) However, when careful precision was not needed, the engineer used a simple clinometer to measure vertical angles.[45]

Gow's surveyors recorded their angles and measurements in a notebook, with the numbers in a table on the left and a sketch of the general appearance on the right. The surveyor measured each "set" of timbers, then sketched them in the notebook in the proper place. In the Butte mines, where square-set timbering was frequently used due to the unstable ground, the regular cubes formed by each set were made to correspond with the graph-paper squares of the surveyor's notebook. The square sets, which were five feet wide and five feet deep in the mine, conveniently translated to quarter-inch squares on a forty-scale map. Once back in the office, both the surveyor and the helper worked up the computed survey numbers independently, to serve as a check against mathematical mistakes, then the correctly computed numbers were entered in a more permanent ledger.[46]

Gow and his mining engineers translated these numbers on to a horizontal plat map. Consistent with precedents established by Brunton, the engineers had a separate map for each horizontal level, with standard symbolic notations for shafts, winzes, and other vertical intersections with the otherwise horizontal level. In keeping with longstanding Western mining practice, each level was named for its depth in the mine, which were commonly one hundred feet apart, so one plat map might show the one hundred level, the next would depict the two hundred level, and so on. On each map, the features of each level, including drifts, crosscuts, laterals, and raises, were numbered sequentially (both on the map and in the mine) with numbers that started with the number of the level. So on the two hundred level, you might find drift 201, lateral 202, and crosscut 203. The engineers also recorded these numbers in a separate book, with a small amount of descriptive information about each of the numbered spaces on each level.[47]

That took care of the levels—the horizontal passageways that allow miners and material to move throughout the underground. But what of the stopes—the working places where the rich veins were mined out? Gow related that the stopes were surveyed and measured once a month. On books of graph paper kept in an underground office, sets of timbers were marked in squares as the mining progressed, and dates were added to indicate when the mining took place. Counting sets made it easy for engineers to calculate the volume of rock removed from each stope, using a set of precalculated tables that took advantage of the fact that each set was the same cubic volume. This information was transferred regularly to a set of similar stope books kept in the mine office building. From this information engineers made additional maps twice a year, longitudinal projections of each lode or vein, which helped mining geologists figure out available ore reserves in the mine. These projections were made along a plane that represented in an average way the sheet of the vein as it cut through the earth, so that these special maps gave the clearest picture of the vein itself, and the workings along that vein on the various levels underground. Each year's work was put on the projection in a different color. Despite the heavily visual character of the work Gow described, it is curious that the monthly reports he made were compilations of data in tables—he did not provide his superiors with maps.[48]

Once the engineers had made their underground maps, which served as the base of Brunton and Winchell's system, geologists went below to fill the maps with geological data. The generation of all this geological information underground was significantly helped by another Brunton invention, one for which geologists today recognize his name. Frustrated by the bulk of traditional underground surveying equipment, so the story goes, Brunton developed a small folding device he called a pocket transit, which he initially patented in 1894.[49] His ingenious device, which weighed less than half a pound, was a precisely made measuring instrument that operated on the same trigonometric principles as ordinary surveying transits. One of the inventor's key insights was that he could eliminate the bulky tripod by utilizing the user's body instead of a tripod as the holding apparatus, as shown in figure 3.1. When the mining engineer held the instrument against his body carefully and correctly, the complex sightlines allowed the pocket transit to work.

The Brunton pocket transit's most important feature was that it was fast and "good enough." The device broke no new ground in terms of precise measurements, and indeed was generally not relied on for careful, complex surveys of the underground. However, the speed with which it could be used allowed work that needed to be only good enough to be conducted far more rapidly than before. Or, inversely, engineers and geologists could do rapid work precisely enough with the pocket transit that it was usable in cases where the very highest degree of precision was not needed. Thus, the Brunton pocket transit combined speed and accuracy in a way that created a significantly greater throughput of usable information.

The Brunton pocket transit was doubtlessly a key technological innovation that permitted Brunton's system of geological mapping to be

Figure 3.1. David W. Brunton's innovative pocket transit dispensed with a tripod by instead requiring the user to hold it tightly against his midsection with both hands, craning his neck downward to peer into the device. Here, it is being used to measure a course or horizontal angle. When held properly, the sight-lines worked and the user received usable results. The lack of a tripod made using the pocket transit an easy way to achieve fast results with rough accuracy, and this was key to collecting the extensive geological information at the heart of the Anaconda System of mapping. From Dunbar D. Scott and others, *The Evolution of Mine-Surveying Instruments* (New York: American Institute of Mining Engineers, 1902), 90.

sustainable over the long run. Mining engineers would make the precise maps of the workings with their super-accurate, but slow transits. Then, Brunton's geologists could use their pocket transits to generate a volume of precise-enough geological data to fill in the level maps. Gathering all the geological minutiae called for by Brunton's mapping system would have been difficult and slow without the pocket transit. Conversely, the imprecision of the pocket transit was made less relevant in the context of short-range underground use for geological mapping, rather than for true surveying.

A now-fragile set of maps used by Winchell in October 1901 as part of this cycle of geological mapping at Butte reveals the messy and contingent process of gathering and interpreting geological data underground.[50] A portion of one of these maps is shown in figure 3.2. The maps themselves are lightly printed positive brownprints, made from the engineering maps of the levels. Winchell folded each of them into a pocket-sized packet about the size of a legal envelope. They bear stains and marks—some seem to be candle wax, others might be acidic or dirty mine waters that constantly dripped on miners underneath Butte—that vividly suggest underground conditions.

The maps are covered in pencil notations as well as blue and red marks. In establishing a mapping tradition that would last decades at Anaconda, blue represented faults and red depicted ore-bearing veins.[51] Given that

Figure 3.2. In 1901 Amalgamated geologist Horace V. Winchell made this temporary map, one of a small number of such maps that have survived. It is a light brownprint of the regular mine map, which Winchell is using to record underground geological phenomena. The map was folded pocket-size, and bears marks from candle wax or mine water, in addition to Winchell's extensive notes in red, blue, and pencil, the first indicating vein matter, the second faulting, and the last his own ideas and other measurements. The notations at the bottom indicate Winchell recording ideas about what he has seen: "This can be considered one large lode, made up of many small veins + veinlets, running in almost all directions + having different dips." Map 7, Folder 2, Collection +83, Newton Horace Winchell and Family Papers, Minnesota Historical Society, St. Paul, MN.

the engineering map he was working from had already been carefully surveyed and measured in the typical way, Winchell could dispense with the normal apparatus of surveying (survey stations, distances, and the like) and concentrate on the geology as it was exposed in the drifts. Even small exposures of vein matter were marked in red and measured—some only inches wide—not because they were worth mining, but because they might hold bigger geological clues. On the maps Winchell's blue-colored faults included rough measurements of the degree and direction of their pitch. He can be seen forming hypotheses and testing them, such as when he writes "Foot wall of vein? No" or "Can This be The Rarus fault?" He summarizes his own growing understanding on the part of the map shown in figure 3.2, declaring, "This can be considered one large lode, made up of many small veins + veinlets, running in almost all directions + having different dips." Winchell also notes the geological composition of the rocks exposed in the drifts, from clay to granite to quartz, as well as ore materials written both chemically ("$CuSO^4$") and by name. He made notes to suggest further development—"Should follow this E[ast]"—and at least once, Winchell revealed his inner geological passion among the technical notations, describing the mineral in a vein: "Enargite (beautiful) in filling."[52] Presumably, Winchell's notes on these maps would have been distilled into reliable geological knowledge and inscribed on the primary geological maps kept safely in the Amalgamated office, rendering these rough drafts obsolete and candidates for the garbage.[53] That these rare specimens survived is a happy accident, for they give a clear glimpse at the messy process of mobilizing the visual culture of mining to capture and wield geological information about Butte's complex underground.

With precise foundational mine maps, and extensive geological information inscribed on them, the final question was how to use these visual representations to change underground mining practices in such a way as to bring greater prosperity or lower risk to the company. Assistant Geologist Frank A. Linforth provided several examples of the application of geological information to underground practices at Anaconda in a paper about the company's geological practices that he presented in 1913 to the AIME when Butte hosted the professional group. One map he presented is seen in figure 3.3. This map shows the final results, but the process of achieving those results involved skilled miners as well as the involvement of the mining geologists.[54]

To understand the process by which geological information and mining excavations created Linforth's map, first imagine the map is blank—solid rock underground, containing unknown mineral wealth or an expanse of valueless granite. As miners in search of ore carved out the underground tunnels, tunnels seem to grow on the map as well. In this case (figure 3.3),

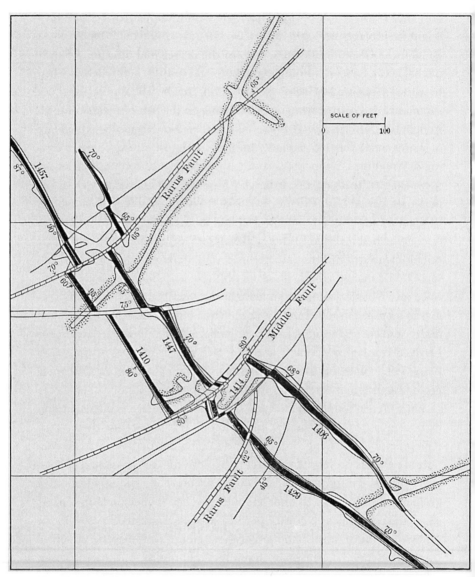

Figure 3.3. This map, showing a portion of the 1,400-foot level of the Mountain View Mine in Butte, was presented to the AIME by Anaconda geologist Frank Linforth. It helps demonstrate how the company could use its detailed knowledge of underground conditions elsewhere in the mine to properly interpret newly encountered geological phenomena. From Frank A. Linforth, "Applied Geology in the Butte Mines," *Transactions of the American Institute of Mining Engineers* 46 (1914): 119.

miners were excavating a tunnel near the Rarus Fault, beginning in the top center of the diagram and moving down and to the left. None of the other tunnels or geological structures shown here had been created or identified yet. The Rarus Fault was an important geological feature in Butte that cut across many veins, and the miners were likely trying to see what else the Rarus might have encountered. Here, they tunneled near but not directly on the fault, because faulted ground requires more support and maintenance. Sure enough, they intersected a vein, labeled 1447 here, by crossing the vein like a T. Heading to their right, back up the map, they followed or "drifted" along the vein until it was cut off by the Rarus Fault. They knew this fault's movement well, so the miners pushed past the fault, cut to the right, and picked up the vein again. After following this vein to the end, they turned around, retracing their steps to the place their tunnel first crossed the vein, and explored the other side (moving down and to the right on the map). In just a short distance, their vein was cut by an unknown fault.

In selective underground mining such as this, miners usually tried to follow the vein, drilling and blasting so that the drift or tunnel they were creating always had the vein in sight at the working face. In this kind of mining, a geological fault is a pain in the neck, because the vein gets cut off and moved—the miners would come back from setting the latest round of dynamite to find the vein had moved or was simply gone. Sometimes it had not moved much, and the miners would have little trouble excavating until it was found. Sometimes the movement of a major fault was well known, as it was here with their first encounter with the Rarus Fault, and it would be easy to calculate the required distance and pick up the vein again on the other side of the fault. Occasionally, there were clues in the rocks, and skilled miners used these along with hunches based on long experience to guess where to blast next to find the resumption of the vein. In the complicated and heavily faulted Butte underground, this sort of hunt was commonplace, and it is easy to understand why Butte miners considered themselves among the world's best.

Tracing the miners' progress down drift 1447, it is easy to see them exercise judgment and skill to keep the vein in sight. The vein matter is represented on the map with a solid black line, and the outer boundary of the tunnel they were making is the thin black line on each side of the dark vein. The new fault had moved their vein, leaving perhaps just a little tiny clue off to the left of the working face. Here, the miners opened up their tunnel by veering left and widening the excavation a little, and were able to find the vein again. This time, the vein was well defined and fairly straight for perhaps more than a hundred feet, until the miners encountered another fault, which cut and moved the vein again.

Next came some confusing, extensively faulted ground. The miners pushed their excavation forward to the left and right, looking for the vein. They soon found another small segment on the left, but that was chopped off again. Exploration to the right paid off next, rewarding them with a short piece of the vein. A small fault displaced the vein to the left once more, but not enough for the miners to lose sight of it. Another fault in rapid succession moved the vein again, and the miners moved cautiously forward and left, where they encountered the vein once more and resumed drifting along the tunnel labeled 1429, hitting and pushing through a localized branch of the Rarus Fault without losing sight of the vein. Had their progress to date been drawn on the map by itself, we would see a single vein proceeding from the upper left toward the lower right, chopped in segments near the middle, but mined out continuously thanks to the skill and judgment of the miners.

Now the geologists brought their maps and expertise to bear on the issue. In accordance with Anaconda practice, they followed the miners almost continuously, recording every geological detail and taking samples of rock to be analyzed. They noted the direction and character of each of the faults the miners passed, changes in the rock, and other information, then brought these data points out of the mine and plotted them on their maps. As the miners proceeded on after the struggles in the middle, the geologists pored over the question. Correlating the exposures they saw here on the 1,400-foot level of the Mountain View with what they knew elsewhere in the mine, they realized they had seen the fault—the one that gave the miners trouble in the middle—before. (Unlike the Rarus Fault, which was a major geological phenomenon that manifested itself in many mines, the newly named Middle Fault was more localized.) But wait— elsewhere in the mine the Middle Fault had a big movement to the left, relative to the direction of the drift the miners had been digging, whereas here it had a moderate throw to the right, which made no sense. The geologists either got the fault wrong, which seemed unlikely given the explanatory power of their careful mapping, or the miners had inadvertently stumbled on a different, parallel vein.

If that was true, a short crosscut tunnel perpendicular to the drift should find the parallel vein. Acting on directions from the geologists, the miners retraced their steps and drove crosscut 1414 to the left, just after the Middle Fault, and found the parallel vein! Retracing their steps farther, to a point just before they had encountered the Middle Fault, they crosscut to the right, and found the parallel vein again! Now that the parallel veins had been found on both sides of the Middle Fault, the miners proceeded to excavate the new veins (drifts 1406 and 1410 on the final map, figure 3.3), and the geologists verified that the Middle Fault now showed movement of a size and direction consistent with indications elsewhere on their maps. Now that they knew the size and structure of the parallel veins, they crosscut to find vein 1457, above the main Rarus Fault, as well. Thus concluded one example of many where the geological information collected by Anaconda mining geologists, inscribed on their maps and sections, helped them visualize the underground in ways that resulted directly in more production and profits for the company.[55]

At the conclusion of Linforth's description to his audience of AIME colleagues, B. H. Dunshee, a mining engineer of high rank for Anaconda, stood up. He gave an impromptu testament to the "exceedingly valuable" geological work from the perspective of the workers mining underground:

> Many times in mining underground we come to a question that is obscure; we don't know exactly where we are or what to do. Instead of going at it blindly and perhaps wasting time and money searching for the continuation of the vein, we merely go to the Geological Department, and the geologists come to our assistance, and almost every time their conclusions are correct; the work they do is intelligent, and again I want to testify to the value of the accurate maps of the veins and fault systems as worked out by our Geological Department.[56]

If the creation and growth of Anaconda's geological department was the fruit of a peculiar set of circumstances, embedded inextricably in the legal and geological context of the time and place, its ascension to greater authority within the Anaconda organization was due in part to the power granted to mining geologists by the visual culture of mining. At Anaconda, the geologists worked assiduously to capture and control the underground with their visual tools, and these tools, in turn, placed predictive power in the hands of the mining geologists.

Recall that in the early decades of American industrial mining, miners viewed geologists with suspicion, perhaps because there was little obvious connection between the sort of work geologists did and the sort of information useful for finding ore underground. A geologist could not see any farther into solid ground than could a practical miner, as the saying went.[57] For millennia, miners had skillfully followed veins and needed no help from geologists to do so. (Indeed, an older term for vein is "lead," because such a vein would guide miners to mineral riches.) Reflecting on this era, Reno Sales observed, "[O]re was plentiful everywhere; consequently to the mine foreman there appeared to be little need for geological advice."[58]

At Butte, however, as the mines were pushed deeper and the geological challenges compounded, the great compilation of geological data about those complex spaces became useful. Brunton and Winchell discovered that their new semitransparent geological maps, which combined engineering information with geological information in a multidimensional yet up-to-date way, attracted attention that either type of information alone did not receive. "Under the old system the mine superintendents and foremen took but little interest in the geological maps," remembered Brunton, "but as soon as the new system was introduced we found them spending a considerable portion of their time each day studying these maps and drawing inferences from them."[59] Winchell described their use by the mining men of the Amalgamated properties in 1906 even more emphatically, noting that the power of the maps had granted influence to their makers. "The mine Superintendents and Managers consult [the geological maps] almost daily, and have come to depend on them almost entirely, although they formerly used to bare plans in the engineer's offices. Indeed the more important mine developments are *now* invariably made in consultation with and under the approval of the geological department."[60]

Looking back during an interview, mining journalist T. A. Rickard compared Brunton's map system to a glass model, because it helped viewers visualize the underground. Brunton demurred: "Yes, but I think that maps of this kind are more convenient and useful than any model I have ever seen."[61] Because Brunton's innovative geological maps sacrificed the total multidimensional visualization power of a glass model to retain the flexibility and ease of use of two-dimensional maps, they were more useful to experts to help them direct the everyday business of mining.

Within a decade, the geological department assumed even more responsibility at Anaconda. Reno Sales recalled that up until about 1914, the group focused almost exclusively on mapping the underground geology and advising the engineers with underground development when requested. After 1914, however, "the engineering department was taken over by the geological department," placing Sales, as chief geologist, in charge of the mining engineers. The mining geologists had used their map-driven geological information to claim the authority over underground exploratory work from the mining engineers. "As a matter of fact, cross-cuts and drifts, nothing was run without our permission," noted Sales.[62]

In 1909, David W. Brunton reflected back on the "progress" made by the mining industry from his position as president of the AIME. The first of two dozen areas of progress that he mentioned was the advancement in mine mapping. Brunton noted that, not many years before, most mines had had only two maps—one of the surface and a second, composite map showing all of the underground workings. Now, he pointed out, most mines augmented these with maps of stopes and maps that recorded assay values, and some companies further recorded underground geology on additional horizontal and vertical maps. "The great advantages of such plans and sections cannot be overestimated. They not only show at a glance the tonnage and value of the ore in sight, but also afford a guide for development-work, whereas the old-fashioned maps were nothing but a record of the work performed, and were practically useless for any other purpose." Brunton also argued that the "improvement" in mapping had caused another change, "the importance of which is just beginning to be recognized." This was the practice of large Western mining companies hiring economic geologists to unravel the geological complexities of their mines and provide advice "of inestimable value" as to the direction of exploration and development work.[63] Brunton modestly failed to mention that his own hiring of Horace V. Winchell at Amalgamated helped show the value such geological experts could have in conjunction with more advanced maps.[64]

Brunton was recognized widely for his efforts, and the innovative geological mapping scheme at Anaconda was certainly an important part of his contributions to the art of mining. In 1927 the AIME awarded Brunton the first gold medal for distinguished achievement in mining. In an article in the AIME's magazine describing his accomplishments, the editor described Brunton's Anaconda work as "the beginning of the system, now so generally followed by large companies, of employing mine geologists" on a continuous basis to record geological data on mine maps and use that data as the foundation for further decisions. Importantly, the editor highlighted how Brunton's work changed the relationship between geologists and those who directed work underground in large mines. "Not the least notable feature of this work was the way it was 'sold to' the superintendents and mine foremen through the mapping of the geology on superimposed transparent sheets . . . with the result that geologists and operating officials were soon working in complete harmony."[65]

Anaconda's efforts in mining geology earned the company widespread recognition for its pioneering work. "The first geological department of any importance established by a mining company—and perhaps the earliest on any scale—was that of the Anaconda company at Butte," noted geologist L. C. Graton in 1947. The company's large scale of operations and commitment to mining geology made it a sort of training ground for young mining geologists, who carried what they learned—"in particular, a standardized and effective method of geological mapping underground"— to mining companies across the globe.[66]

These pioneering efforts were a result of new ways of mobilizing and utilizing the visual culture of mining in everyday work. Reflecting on his department in 1929, Anaconda chief geologist Reno Sales noted how important an engineering education, and the consequent inculcation of visual thinking that accompanied that education, was to the work they did. Whether the job was geologist, mining engineer, or assay sampler, to work for Sales at Anaconda required thinking like an engineer. "Experience has shown that men trained in professions other than engineering do not get a satisfactory grasp of the problems involved in mine operations," he noted. Then, he let the numbers speak for themselves—of 133 people working in his department, only a single one had not attended college as an engineer.[67] Clearly, for Sales the ability to wield visual representations in a complex system was of paramount importance.

The underground maps made by mining engineers foregrounded the engineered underground. These maps were built to highlight what the engineers built—tunnels, drifts, gangways, and stopes. In the hands of the mining geologists, these maps were turned to a different purpose. The tunnels and drifts became the places were geological phenomena were glimpsed. The centers of the stories told by the geological maps were situated in the blank spots—the unexcavated, unknown places beyond tunnel walls. The mining geologists collected every glimpse of geological information they were permitted by existing excavations to see, and over time these little bits aggregated, on their maps, into larger pictures of the complex Butte underground that allowed the geologists to make good guesses about what miners would find in unexcavated places. By turning their underground maps around, seeing not the built world of the engineers but the geology that likely rested in the rock beyond view, mining geologists turned the visual culture of mining to work in new and powerful ways.

PART II

Mine Models

chapter 4

Modeling the Underground
in Three Dimensions

The two-dimensional maps described in the previous chapters were the most common visual representations associated with the mining industry in the late nineteenth and early twentieth centuries. Though these maps were important and could be beautiful, the most *spectacular* elements of the visual culture of mining were certainly the many models that attempted to display the invisible underground in three dimensions. Models first appeared in the American mining industry in the decades after the Civil War, but only became common in America after the turn of the twentieth century. Three-dimensional models were some of the most powerful representations of the underground, but access to their visual power came at a price. These models were expensive, hard to handle, and difficult to update with additional information. As a consequence, models were used only when more-traditional visual technologies, such as maps or drawings, were considered ineffective. The special power of models stemmed from their ability to communicate information to nonengineers in a way that was difficult for maps to do. Models were used mostly in classrooms, courtrooms, and exhibitions—situations where their inflexibility was not a detriment, and their particular powers were of paramount importance.

This chapter will explore technical models in the American mining industry during its era of professionalization. This will include how they were made, their various forms, and a broad survey of their uses in classrooms and courtrooms. Chapter 5 will examine the story of a particular model's use in a mining lawsuit. The term *technical models* perhaps deserves further explanation. This term is used here to describe three-dimensional representations of underground spaces that were constructed for the primary purpose of conveying measurable information about the underground, testing mechanisms, showing relationships between things, and so on. These technical models were not originally made to be put in a museum, even if some did end up there eventually. Technical models stand in contrast to lifelike display models, which are found primarily in museums and expositions. The purpose of display models is to portray a broader, less technically sophisticated image of mining work; these models were frequently directed at a general public audience. More will be said about display models, with their roots in expositions and museums, in chapter 6. There was some overlap between display models and technical models, but it is useful to make a distinction because it can help clarify issues of audience, authorship, and style. It may be helpful to compare technical models, part of the visual culture of mining, with other models used in scientific inquiry.[1]

The term *model* has multiple meanings. One way to look at a model is that it is a system of understanding that simplifies inputs and predicts outcomes, but that does not necessarily have a material representation. These *theoretical models,* such as those used to predict outcomes in economics, did not play a discernible role in the nineteenth- and early-twentieth-century mining industry and would not have been represented as technical models. Other scientific models, however, are material artifacts intended to depict physical relationships and phenomena in a scale different from that of the original, and thus bear some resemblance to underground mine models. The mine models discussed in this chapter were material artifacts that used three dimensions (instead of a map's two) to represent underground spaces. While historians have long paid attention to the *concepts* that scientific models represented, the physical makeup of models, that constrains the way they can convey those concepts, has not been well studied.[2] Mine models provide an excellent opportunity for this sort of historical analysis.

Mine models functioned differently from many of the material models previously described by historians of science. For example, think of the three-dimensional ball-and-stick models of atomic particles that were used in twentieth-century chemical sciences to build molecular structures.

(These are no doubt familiar to every person who has endured basic-level chemistry.) These molecular models aided scientists in their work by being manipulable, flexible objects that themselves "embody, rather than imply, the spatial relationship of the molecule's components." Put another way, the "models mimic, mechanically, some of the important physical properties attributed to molecules."[3] Scientists could test hypotheses and experiment by manipulating the model. By simply rearranging the balls and sticks, the model helps them understand if a hypothesis might possibly be true. If the pieces of the model did not fit, the molecules they modeled would not work that way either. As a result, molecular models could help scientists solve problems mechanically that were analytically out of reach in the pre-supercomputer era. Thus molecular models could shape scientific thinking.[4]

Mine models, in contrast, functioned simply as representations of the underground. In the case of such a representation, the manipulation of the information borne by a mine model took place only in the viewer's "mind's eye," as was the case with engineering drawings and mine maps.[5] Rearranging the sheets of glass in a glass plate model would have yielded nothing but confusion. In other words, mine models were only suggestive; by showing what was known, they helped the viewer imagine the contents of the unknown. Molecular models, by contrast, could take the physical form (in facsimile) of an unknown thing and allow the scientist to manipulate it directly.

A related tradition of model building, called *study models* in historian Eugene Ferguson's formulation, had a long history in engineering, dating to before the widespread adoption of engineering drawings, but was more rare in mining engineering.[6] Nonetheless, some examples of mining-related study models have survived. A wooden model of a vertical shaft compartment in the C&H collections at the Keweenaw National Historical Park might have been useful to help plan the internal structure of a shaft before it was constructed.[7] Similarly, the Amalgamated Copper Company (later known as Anaconda) made a wooden model that showed the unusual timbering system used in the Pennsylvania Shaft, in Butte, Montana, due to the difficult "heavy" ground through which the shaft passed.[8] In the early 1930s Philip Bucky, a young professor of mining at Columbia, conducted some tests to see if scale models of underground structures could be used to predict material failure when subjected to appropriate forces by being whirled in a centrifuge. He thought his results might be promising, but the complexity and detail of his methods, combined with a lack of precise instruments, ensured that little ever resulted from the idea.[9] An example of a study model and its full-size analogue can be found in the Missouri lead belt. A photograph shows a small model of a platform made of boards and wires, made by the St. Joe Lead Co. of Missouri, circa 1950s.[10] The real-life platforms were intended to allow miners to drill into the roof of a mine without support from below or from either side. Full-size examples of this homemade platform system can be seen in situ in the enormous underground excavations of the Bonne Terre Mine, which the St. Joe Lead Co. in Bonne Terre, Missouri, operated until its closure in 1962.

In these cases, in the manner of scientific models, study models could be manipulated to determine whether the relationships expressed by the model would work at full scale. Nonetheless, the majority of technical models used in a mining context were not study models designed to educate their users through manual manipulation or to help solve technical problems at a smaller scale, but instead were intended to encourage viewers to imagine underground mines and draw their own conclusions about geological or mining problems.

Engineers in the late nineteenth and early twentieth centuries did not generally use technical mine models as part of their everyday practice. "[C]omparatively few mine models are made for mine use," noted mining engineer C. L. Severy in 1912.[11] Instead, technical models were reserved for special circumstances like lawsuits. However, despite their importance in lawsuits, technical models were rarely used in day-to-day engineering work in this period because updating even the easiest-to-use models was a tedious chore. One model maker lamented in 1917 that it was difficult to convince the engineers of the worth of models. "When it comes to taking such a model apart for the purpose of making addition[s], the engineers and draftsmen are often inclined to lose interest and neglect it, especially as the practical part of their work is better served by the mine drawings on a working scale, which furnish them a much more intimate knowledge of underground conditions than a model. To the operating staff at the mine, a model is generally considered a clever bit of ingenuity, but of little practical value."[12] These "clever bits of ingenuity" generally took one of a handful of common forms, and could be made of almost any material. Three types of technical models—block models, sectional models, and negative or skeleton models—became most common, and each of these forms will be described in greater detail below. In contrast to technical models, life-like display models depicted mining naturalistically, often portrayed only the surface of the earth instead of underground workings, and were especially popular for museums and expositions. Because these were meant primarily for the public, rather than for mining industry audiences, they had a different history, as we will describe in chapter 6.

There are three broad categories of forms of technical mine models, though there was no historical consensus on the appropriate terms for them. *Block models* were made of solid materials and ordinarily depicted underground geology. *Sectional models,* which included glass plate models, showed mine workings and geology in semitransparent slices that were either horizontal or vertical; these were the most popular form for underground mine models. *Negative models,* often referred to as skeleton models when they depicted worked-out underground veins, used material to represent space, and space to represent solid rock. Some models combined aspects of multiple forms.

T. S. Harrison and H. C. Zulch, surveying the types of models and their uses in mining lawsuits in 1908, suggested that each of the different forms of models had particular strengths. Sectional models, including glass plate models, were inexpensive to build and were useful for correlating veins and underground workings. Negative, or skeleton models, had the advantage of showing "the entire mine workings in their relative positions at one comprehensive glance."[13] Block models, by contrast, were most useful to depict simplified geological phenomena, such as when the court's attention focused on a single vein.[14]

Block models were solid models that either omitted surface topography or showed a cutaway view of the earth. These block models were derived from an older tradition of geological modeling that predated the period of industrial mining in America. These models were not always as useful in a technical way as were the later glass plate and skeleton models, but were applied to mining geology where appropriate. Their usefulness for directing work was limited in part because of their less-precise construction and small scale, but they were often more aesthetically pleasing than the alternatives, could be less expensive to build, and were better suited to interpretation by nonexperts.

Early examples of geological block models were made by Thomas Sopwith, a British engineer. Sopwith was one of the pioneers of his craft.[15] Sopwith noted in 1834, "Topographical modelling is scarcely either known or practised; and when it is considered what extravagant sums are daily expended in mere trifles, it is surprising that a pursuit combining so much elegant amusement with practical science and utility, should be almost utterly neglected."[16] His 1834 treatise on drawing included instructions for building models that depicted geology. Sopwith advocated imposition of a grid of squares on a geological map, then creation of sections, or vertical depictions of the geology, along every line. These sections were then traced on to pasteboard or thin sheets of copper, and colored in. A series of half-notches were cut at appropriate intervals, so that the pasteboard sections fit together in a hollow grid in an egg crate pattern. Finally, wood or plaster of paris blocks could be carved to fit in the hollow squares to represent the surface topology, but could be removed if the viewer wanted to see the geology below.[17] Several models built by Sopwith using this method were exhibited for years at the Museum of Practical Geology in London.[18]

Sopwith's models attracted the attention of other mining engineers and geologists. Geologist William Buckland persuaded Sopwith to sell a set of twelve of his smaller models for the purpose of teaching geology, and Sopwith and his cousin, John Sopwith, began producing them in 1841.[19] These small square wooden models showed vertical sections on four sides, the surface topography on top, and the underground geology, projected on a horizontal plane, on the bottom. The models were expensive to produce, but demand was steady, and a series of six simplified and updated models was issued in 1875. The 1875 set of six, available for three guineas, contained square models four inches to a side, but larger versions were available for lectures.[20] Both sets were accompanied by an explanatory book that led the reader through a description of the geology depicted on the models much as a lecturer might have done.

The construction of even approximately accurate block or topographic models, such as those Sopwith produced, depended on having reliable information about elevation. If, however, underground elevation information had been collected (or estimated) and plotted on a two-dimensional map in the form of contour lines, then the creation of a three-dimensional model based on that elevation data required some fine craftsmanship but was not an intellectually difficult enterprise. Some, though not all, of the maps produced by the so-called Great Surveys of the 1860s and 1870s used contour lines to indicate elevation, though some of the surveys preferred the older method of using hatchures to indicate vertical relief.[21] However, after the consolidation of the Great Surveys and the formation of the USGS in 1879, contours became the standard means of depicting quantitative elevation data on maps produced by the USGS and state surveys. Several model makers took advantage of this information to create three-dimensional block models that depicted vertical relief.

J. Peter Lesley was the long-time director of the Second Geological Survey of Pennsylvania and professor of geology at the University of Pennsylvania. Over his decades of teaching geology, Lesley experimented with several ways of creating topographic models, and in his role as director of the Pennsylvania survey he served as a mentor to later model builders. Lesley also helped develop and teach the methods of topographical surveying and mapping, using contour lines, that produced maps suitable as the basis of topographical block models.[22] Lesley tried making models using a wide variety of techniques, including those pioneered by Sopwith; he also tried simply carving a block of wood (a block model in the truest sense). However, the method Lesley and those he trained and employed, most notably including E. B. Harden and O. B. Harden, used most often was building up a model by cutting out a series of contours and positioning them on top of one another. Each layer was cut to fit a contour of the map, and the thickness of the material of each layer provided the vertical elevation desired by the model maker. Wax, or other pliable material such as clay, was used to fill in the stair-step hillsides created on the model by the technique; after final shaping, the model was covered with shellac or other preservative materials. The final step consisted in covering the wooden model with plaster to make a negative cast, from which positive copies could be made. Plaster copies would not crack from shrinkage of the wood, and could be individually colored to show geology or other features, but had the disadvantages of being heavy, expensive, and fragile.[23] Model makers could also saw apart such models and paint the geology on the vertical sides.

A significant point of contention in the construction of topographical or relief block models was the degree to which the vertical scale should be exaggerated, if at all. On a model covering a sufficiently large area, even significant topographical features such as mountains could seem insignificant at their true scale, and minor variations on relatively flat land could hardly be noticeable. The use of an exaggerated vertical scale avoided the appearance of distortion to those who were unaccustomed to looking at such models. With a uniform scale, steep portions appeared excessive and flat parts seemed too flat. However, a modest vertical exaggeration was "found to be the best average and convey[ed] the most natural impression to the eye."[24] In summarizing the various topographical modeling techniques published by the late 1880s, model makers John H. Harden and Edward B. Harden noted the "great diversity of opinion" on the topic of exaggerated scales, and thought "the weight of opinion seems to be in favor of a slightly exaggerated scale."[25]

Other model makers, especially Lesley himself, argued strenuously for uniform horizontal and vertical scales. John H. and Edward B. Harden pointed out that they thought exaggeration was particularly unnecessary for large-scale models (that is, those depicting a small enough area for vertical relief to show through naturally on the model). The Hardens were working model makers, however, and keeping in mind the potential demands of clients, pointed out that they avoided exaggeration "where possible."[26] Lesley took a stronger view, befitting his stature as a well-regarded American scientist, and reminiscent of his earlier battle against exaggerated scale in two-dimensional geological sections. Lesley pinned the desire for exaggerated relief on the failings of the model maker:

> The demand for exaggeration in a relief comes from those who will not spend a sufficient amount of time and pains upon the intermediate contour curves, or from those who have not trained themselves in drawing from objects. The habit of exaggerating the relief excuses itself at first on the plea that common people cannot appreciate heights when true to nature, but the fact is that the difficulty is felt by the modeller himself; and when the habit is once formed, it becomes incurable. If a relief-map be not true to nature, what is the good of it?[27]

Lesley could influence model-making practice in part because he was among the few people prior to the 1890s actively involved in making and ordering models. In his position as professor at the University of Pennsylvania, Lesley hired one of the Hardens and delegated his students to help build at least fourteen models to help illustrate geological phenomena. In his role as director of the Pennsylvania Geological Survey, he paid the

Hardens to build models of important regions of the state derived directly from topographic and geological data collected by his survey.[28]

Just as elevation data about the surface of the earth, plotted on topographical maps in the form of contours, could be used to create topographic block models, the development of maps that represented underground elevations with contours also facilitated model making.

The first maps with underground elevation contours were made by Benjamin Smith Lyman in the mid-1860s. Lyman recalled plotting some properties in Virginia using underground contours in 1866 and 1867, and showing off a photograph of one of these maps in 1867 to the American Association for the Advancement of Science. Lyman first published a map featuring underground contours in 1870 (of oil fields near Lahore, India), but the idea was still quite new when he published a description of the technique in 1873.[29]

The first known block-style model that depicted the underground inside a mine directly was also made in 1865, though there is some confusion about the precise details of its genesis. Lesley noted that he made a model of the Plymouth anthracite colliery in 1865 using elevation data that he personally recorded inside the mine. Harden and Harden note that the first model was indeed of Plymouth, and used underground contours from a map made by Lyman.[30] In 1873, using their familiar technique, the model makers made at least one other block model for Lesley showing the underground, to help him illustrate geological structures for his students.[31]

The extensive amount of data collected by the Second Geological Survey of Pennsylvania, under Lesley's direction, spawned several important block models depicting the underground. Charles A. Ashburner, who played a major role in the Second Geological Survey of Pennsylvania, used Lyman's underground contour technique to produce a series of maps and an impressive block model of the Panther Creek coal basin in the anthracite country of Pennsylvania circa 1882; in doing so, he used geological modeling techniques to present information of direct use to those in the mining business.[32] The Panther Creek basin was probably the most geologically complex of the anthracite areas, and Ashburner anticipated that a solid understanding of the extent of the coal at depth could provide useful information to coal entrepreneurs. Ashburner's geological survey team collected elevation measurements of the bottom of the Mammoth Vein as it was exposed in the underground workings in the Panther Creek basin.

Ashburner then created underground elevation contours, fifty feet apart, which he plotted on maps at a scale of eight hundred feet to an inch. Though he published these maps with his report, he also wanted to represent the coal bed in a different way. Ashburner hired John Henry Harden to construct a three-dimensional model using the contour data. Harden used his preferred method for making topographical models to build Ashburner's vision in stacked contours, each representing one hundred feet. Harden created a negative cast from the wood and wax original, which was then used to produce at least one positive cast in plaster of paris. To portray the model in Ashburner's report, a photograph was made of the model, then a lithograph was made from the photograph for duplication of the image.[33] A copy of his model was presented at the New Orleans Exposition of 1884, and later went on display at the United States National Museum by 1891, in the collections of economic geology.[34]

Ashburner was justly proud of his work, which directly coupled earlier geological and topographical modeling traditions with the geological needs of anthracite mining engineers and owners. Ashburner noted, "[T]his was the first time that such an underground model was constructed in this way, and for the purposes for which the Panther Creek model was designed by myself; it has attracted the attention of many of the mining engineers of the anthracite fields, and of eminent professional geologists."[35]

Bennett Brough, the British author of the standard treatise on mine surveying first published in 1888, summed up the importance of Ashburner's model: "Thus, the final model, made in wood and wax, not only formed a graphic representation of the structure of the strata in a highly plicated district, but also proved of great value in the definition of its geological structure, and in the deduction of many conclusions affecting the amount of coal contained in this coal basin, and the proper methods to pursue in its ultimate mining."[36] Block models, whether made in the built-up contour fashion employed by the Hardens and others or simply carved from a solid mass of wood or plaster, could be broadly instructive about topography and geology over a large area. In contrast to the sectional and negative models discussed later in this chapter, block models could present different views by utilizing multiple pieces that fit together; when observers removed those pieces, they would see revealed additional information. One such block model can be seen in figure 4.1. This model, from the

Pottsville, Pennsylvania office of the P&R C&I, likely dates from the nineteenth century and depicts the company's anthracite lands near its Wadesville shaft. This block model depicts the surface topography and the geology of the overlapping anthracite veins worked by the company. A vertical section of the vein structure is shown on its flat sides. Made of wood, the model's pieces are held together by a series of latches, making it possible to unlatch and remove each layer. The model's layers come apart at each vein, and follow the undulating movement of the coal geology. As can be seen in figure 4.1, removing the top piece of the model reveals the Diamond Vein and some of the company's main underground infrastructure in that seam. Removing the second layer reveals the Primrose Vein and its workings, and lifting the third piece shows the Seven Foot Vein, along with a projected gangway along the lower edge of the model. Despite the

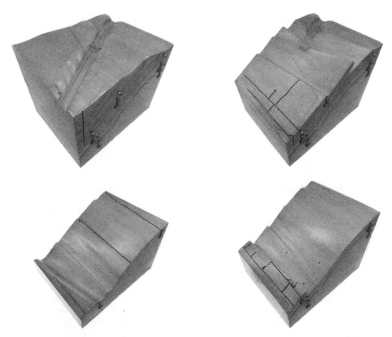

Figure 4.1. This block model, owned by the Philadelphia and Reading Coal and Iron Company, depicted the undulating and overlapping anthracite coal veins near its Wadesville shaft. A vertical section is shown on the model's flat sides. The model consists of four layers so each vein can be lifted off to expose the one below. *Clockwise from the top left:* ground surface, Diamond Vein, Primrose Vein, and the Seven Foot Vein. Model in the collections of the Division of Work and Industry, National Museum of American History, Smithsonian Institution. Photo by Eric Nystrom, 2013.

model's monochromatic simplicity, the intricate layers vividly depict the folded and buckled coal seams with which the company had to contend.[37]

Block models could be built to utilize the vertical joints of the model, where the pieces came together, to represent information and underground features in different ways. The model in figure 4.2 shows a clear example. The planes of the joints could display a projected section—an idealized geological abstraction. They could also follow faults or other natural geological features.

A block model made circa 1898 by the Anaconda Copper Mining Company of Butte, as a defendant's exhibit in its suit with the Colusa-Parrott, has a multipart design that opens to reveal idealized vertical sections. The model comprises three vertical slabs, held together with door hinges, that swing open to show the vein inside; the top is carved and painted to represent the surface workings. The tall model shows four painted sections of the steep vein, one each on the outside faces of the model, and views looking east and west of two sections painted on the insides of the slabs.[38]

Mining engineer David W. Brunton, who was introduced in chapter 3, used images of a large block model of Aspen Mountain, in Colorado, to illustrate an article he wrote about Aspen's geology for the *Engineering and Mining Journal*.[39] The model and geological information he presented stemmed from an apex lawsuit between two of the major mines there. His model was topographically correct on the surface, having been constructed from contours measured twenty feet apart on the mountain. Brunton's block model was built to split apart along a significant fault line, and thus show the interior geology and faults of the mountain. Brunton helped facilitate the later donation of this model to the Smithsonian Institution.[40]

Some block models combined the techniques of displaying sections and workings, in concert with an approach using smaller units that was

Figure 4.2. Block model made of wood, created for a lawsuit in Cripple Creek, Colorado, which shows the surface topology and can come apart to reveal vertical sections. From T. S. Harrison and H. C. Zulch, "Court Maps and Models," *Mines and Minerals* 29 (September 1908): 51.

reminiscent of Sopwith's early geological models. W. I. Evans described a block model of the Copper Cliff nickel mine in Sudbury, Ontario, that had been made in a very modular fashion, designed to open along idealized sections as well as along the workings. The model itself was created from blocks of wood five inches square, each carved to accurately represent the plan and section of the surface topography. Tiny bits of ore and rock from the site were glued on to the blocks to represent the geology. To depict the underground levels, additional blocks were stacked on each other, so removing the top block exposed the level beneath. Evans noted his model was particularly helpful to show the mine to "the shareholders, or intending stock buyers, who could get but a very crude idea of [the mine's] form and extent from plans."[41]

The block model in figure 4.2 shows similar features. This model was created for an apex lawsuit between the Aileen and Mary McKinney claims in the Cripple Creek District of Colorado. In this case the model was built in pieces and cut along the vein, as well as along five vertical sections. Along the vein, existing workings were carved out of the wood and vein rock in place was painted. A series of bolts could be used to hold together only those pieces of the model that were needed to illustrate a particular feature.[42]

Despite, or perhaps because of, the seemingly infinite variations on the form of the basic block model, these technical representations were the oldest and simplest way to display idealized geological information about the underground or surface. The advent of contour maps helped model makers create such block models with greater ease, and the prospect of combining a vertical section in the model, as in the Cripple Creek, Aspen, and Butte model examples mentioned above, helped make block models an effective means of displaying generalized underground information. For displaying underground workings with greater precision, many mining engineers turned to a different form of model, made in sections.

Sectional models, made from sheets of glass or some other transparent material, ultimately became the most common type of three-dimensional model used in the mining industry. Glass was by far the most common material used for these types of models, but they were occasionally made with other materials such as sheets of celluloid.[43] Compared to block models and skeleton models, made of wood, metal, or plaster of paris, sectional models had two distinct advantages: they were inexpensive, and they

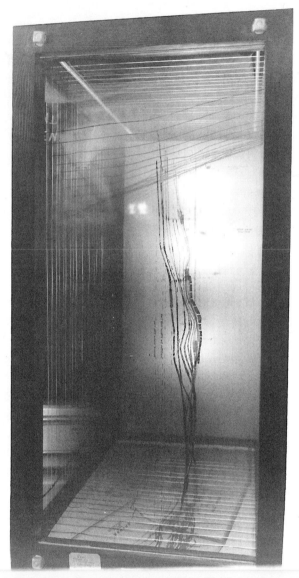

Figure 4.3. Individual two-dimensional vertical sections or horizontal plats were traced onto glass, then the plates arranged in a case that held them a fixed distance from each other. Looking through the plates created a three-dimensional view. This model was made for an Idaho court case in the 1910s. "Defendant's Exhibit 11," Bunker Hill / Pintlar Corporation Collection, Manuscript Group 413, Box 337, folder 5933, Special Collections and Archives, University of Idaho Library, Moscow, ID.

were easy to make and update.[44] Figure 4.3 shows a glass plate model constructed circa 1917 for an Idaho lawsuit.

To build a sectional model, the model maker would paint or draw the underground workings on a series of sheets of glass (such as common window panes). The maker would then insert these sheets into a box-like frame, with rails or notches cut in the sides to hold the glass in the proper positions. The model maker could arrange the sheets vertically, to show sections of the mine, or horizontally, to correspond with the levels of the mine.[45] A top sheet might depict claim lines and other surface features. The spaces between the glass sheets might also be to scale, to indicate the actual space between levels in the mine, or it might be a consistent diminution of the actual vertical scale. If the model was planned for the same scale as that used by the mine's maps, then marking the plates was simple. The modeler simply placed the glass on top of an unrolled map and traced the workings. A slightly more advanced practice was to turn the mine maps over, so that the tracing appeared from below. This was backward when the glass was placed on top of the reversed map and inked in, but it meant that the drawn section was now on the underside of the glass when placed in position in the model. This made the model easier to clean, since there was no worry of wiping off the ink along with dirt on top of the plates.[46] If a different scale was necessary, the work was slightly more complex, but in all cases working in two dimensions enormously simplified the model-making process.

The earliest known example of an American glass plate model was built by Charles T. Healey in 1874. The model depicted the New Almaden quicksilver (mercury) mine in California, and was used to help understand the complexities of the vein system, with the eventual hope of finding new ore. The New Almaden, named to capture some of the glory of the world's most famous and productive mercury mine in Spain, was claimed and brought into production in 1845 by Andreas Castillo, a Mexican army officer.[47] Mercury's major use was as a key ingredient in the refining of gold and silver ores through amalgamation. Demand for mercury from Mexico and China was already high enough to justify the rapid development of New Almaden from its discovery in 1845. The advent of the California Gold Rush in 1848–1849, and the development of the Comstock Lode's gold and silver ores beginning in 1859, both opened large domestic markets for California mercury that stimulated production.[48] At New Almaden, as in other mercury mines, the cinnabar, or mercury ore, seemed to occur only in rich pockets, most of them unconnected. Sometimes old pockets would lead to new ones, and sometimes miners seemed to get lucky, but the search for fresh deposits was even more frustrating at New Almaden than at most mines because there did not seem to be a clear system of veins for the miners to follow. New Almaden's peak production (before 1888) was in 1865, when the mine produced 47,194 flasks of mercury. (A flask, as usually measured in the United States, is approximately 76.5 pounds of mercury.) However, by the end of that decade and the beginning of the 1870s, known ore reserves had been largely exhausted and production was declining. The year 1874 saw the lowest recorded production for New Almaden, with only 9,084 flasks for the year, a decrease of some two thousand from the year before.[49] New Almaden's decreased output drove the price of mercury up significantly in 1874–1875, which prompted the frantic search for new deposits at New Almaden and other mines in California. Adding to the price pressure on refined mercury was the increased demand of the mills on the Comstock Lode. In 1873 the Consolidated Virginia Mine found the rich gold and silver deposit immortalized as the Big Bonanza, but mercury was a vital ingredient in the transformation of Consolidated Virginia ore into a salable metal.[50] Thus, in 1874 the pressure to discover new orebodies at the New Almaden was intense—production had fallen, little ore was in sight, and the demand for mercury was huge. The mines' operators embarked on a major mapping and exploration effort, but it was not clear if it would yield the ore they needed. Thus, the creation of a glass model of the New Almaden Mine in 1874 should best be understood as an attempt to rearrange existing data in a format that might permit engineers to perhaps divine some new, previously overlooked relationship between the orebodies that might lead to new deposits.

Charles T. Healey, the builder of the model, spent most of his life as a civil engineer employed by the government. Healey was a surveyor for New Almaden, but was let go as an economy measure sometime after building the model. He found an engineering job at a nearby mercury mine, the Guadalupe, and afterward conducted an independent engineering practice in San Francisco.[51] Healey's model was exhibited in 1874 at the Pavilion in San Francisco, which was an annual trade show, sponsored by the Mechanics' Institute of the City of San Francisco; the event facilitated the city's trade in technology, especially mining technology.[52] "We believe this is the first attempt of this mode of mine delineation," wrote the editor of the *Mining and Scientific Press*. He sounded thunderstruck: "It is not only an exquisite work of art, but it constitutes the most palpable, truthful and comprehensive work of representing the underground workings of a mine which has ever been devised. No description can do it justice; it must be seen to be appreciated. The whole mine . . . [is shown] just as it would appear to a clairvoyant who might possess the power, while standing upon the lower portion of the surface of the mine, to look directly through the earth and rock, and view collectively the entire underground work."[53] The model took a form that would still be recognizable in glass plate models half a century later. It consisted of twenty-six glass sheets, each approximately twenty-six inches long and eleven inches wide, set vertically in a notched frame that held the plates exactly an inch apart. The frame was mounted on a pivot to allow the model to be rotated before the viewer. Each glass plate represented a vertical slice of the underground geology. The modeler painted the surface topography, complete with buildings and trees, then painted representations of the insides of the mountain—misty gray for the ordinary rock; dark black for exploration tunnels that had come up empty; and vivid red, the color of cinnabar, for the chambers where ore pockets had been found and worked out. The vertical sections were based on a scale of one hundred feet to an inch, both horizontally and vertically, so when combined an inch apart in the frame the glass plates created a model of consistent scale.

The model's utility to cinnabar prospectors was immediately obvious to the *Mining and Scientific Press*. The separated orebodies of the New Almaden Mine were clear, but a viewer slowly turning the model on its pivot could see the deposits to have "a most striking general conformity" with the slope of the surface. (The initial discoveries had been on the top of the aptly named Mine Hill.) This correlation, made possible only by the three-dimensional view of Healey's model, "may afford some important hints for prospecting elsewhere among the quicksilver deposits which are now being so largely sought for in various portions of the State." The editor further recognized the educational value of such models in training mining engineers, and strongly suggested that the mining program at the University of California should acquire Healey's model and possibly others built the same way.[54]

Assessing the impact of the 1874 glass model of the New Almaden Mine is difficult. Though at least a few European models utilizing glass plates had been constructed more than a decade earlier, Healey's model seems to have been the first glass plate model mentioned in the American mining press, and its exhibition at the Pavilion would have brought many people, from both mining and other industries, in contact with it. For the mine itself, it seems likely that the comprehensive mapping program carried out in the 1870s and the higher price of mercury contributed more to the revitalization of the New Almaden's production than did one small model. In fact, the better maps were probably necessary for the creation of the model at all. However, the model may have made it easier to understand the complex geology of New Almaden. In his USGS monograph produced in 1888, geologist George F. Becker described a system of two roughly parallel fissures that seemed to have helped form and also constrained the deposits. The fissures were difficult to represent on a map or a vertical section, cautioned Becker, because they curved through the mountain. "Could one but represent the fissures by contours, the entire structure would be shown in three dimensions and would not be ambiguous."[55] By portraying the orebodies in three-dimensional space, Healey's 1874 glass plate model provided the clarified view so helpful in understanding the underground landscape of the New Almaden Mine.

Even after Healey left the company, models remained an important way of understanding the geology of New Almaden. Mining engineer Hennen Jennings, while at New Almaden after 1877, met Edward Benjamin, who was "at that time model-making."[56] In 1888 James B. Randol offered the United States National Museum a model of New Almaden made along these lines. When a museum administrator asked Frederic P. Dewey, curator of economic geology and metallurgy at the museum, if he wanted the model, Dewey replied, "Most certainly Yes!" Randol shipped the model to the museum and by 1891 it was on display. Dewey's 1891 catalog of the economic geology collections at the United States National Museum noted a "fine glass model" of New Almaden among their displays. This model also comprised twenty-six glass plates, but Dewey listed the plates as representing fifty feet from each other, rather than the hundred feet of the Healey model.[57]

After the creation of the New Almaden model in 1874, glass plate sectional models slowly became the most common type of model of underground mines, even though the models themselves were still relatively rare before the turn of the century. In 1877 Healey helped spread information about glass mine models among mining engineers when he helped another engineer create a similar model for the litigation between the Richmond Mine and the Eureka Consolidated Mine, both in the Eureka District of Nevada.

The *Eureka-Richmond* case attracted a large amount of attention in the mining press because of the unusual geology, high monetary stakes, and extensive participation by expert witnesses, as well as the "elaborate and extensive maps and models."[58] The case was tried in San Francisco, despite the location of the disputed mines in Nevada, which also increased the visibility of the proceedings. Each side in the case made a sectional model out of glass. The Richmond Mining Company created a large model, six feet long, four feet wide, and three feet high, at a scale of forty feet to the inch. To construct this model, mining engineer Nathaniel Westcoat traced the primary map of the mine on glass plates, and inserted small vertical pieces of glass where appropriate to show the continuity of workings and orebodies between levels. Westcoat portrayed the workings on each level in a different hue, which was common practice on a composite mine map but ensured the model would be a cacophony of color.[59] Westcoat was assisted in his model making by Charles T. Healey, who had made the earlier New Almaden model. The San Francisco–based *Mining and Scientific Press* noted the connection directly: "The device is similar to the one showing a model of the New Almaden mine, exhibited at the Mechanics' Institute fair a few years ago. That, however, was on a smaller scale, although made in the same way."[60]

The Eureka Consolidated Mining Company's model used vertical sections of glass, traced from vertical sections submitted as part of an expert's affidavit. However, the accuracy of the model was quickly called into question when a sharp-eyed opposition attorney noticed an embarrassing model-making mistake. He realized the Eureka's model showed a neighboring mine to the east, instead of its proper place to the west. The engineer had inserted the plates in the wrong order. Although mine models could carry significant persuasive weight in deciding cases, poorly designed models might negatively affect the outcome of a case.[61]

The fame of glass models was further spread through their use in a trial in Leadville, Colorado. In 1878 mining engineer Arthur D. Foote made a small glass model of the vein in the Iron Silver Mine at Leadville to support his company's side in a lawsuit. In addition to the model, Foote conducted extensive surveys and made many maps of the mine. The earlier *Eureka-Richmond* case as well as the Iron Silver and other Leadville cases likely spread knowledge of glass models among mining engineers. It was clear that the general public, at least, was unfamiliar with such models. Foote's wife wrote in her memoirs, "His glass model of the vein scored a hit with the jury who admired it like a toy; for awhile it became a quite celebrated little toy."[62]

Foote's earlier work as a mining engineer, before he built the glass model at Leadville, was for the New Almaden Mine. He also mapped and surveyed that property, beginning employment in 1875 or 1876. It is likely that Foote became acquainted with Healey's model of New Almaden during his work there, and then built the Leadville model drawing on his previous experience. Foote got his job at New Almaden after Healey refused to return to work for the manager James B. Randol.[63]

The ease of constructing sectional models of glass made models of this type by far the most common in the early twentieth century mining industry. In an article describing the operations of the Phoenix Mine in British Columbia, mining engineer C. M. Campbell noted that the company kept a horizontal glass model of the mine's stopes up to date. This work was made easier because the stope maps and the model made from them were on the same scale, so modelers could trace new features directly onto the glass sheets. Additionally, Campbell reported the company was considering making a vertical sectional model by tracing, in similar fashion, the vertical geological sections already prepared by engineers. The models complemented day-to-day work at the mine but were not integral to it, reported Campbell. "This arrangement does not need to be referred to often, but when it is needed it is found to be of very considerable help."[64]

A rarer form of sectional model featured glass plates carved to represent the workings, rather than utilizing drawn or inked information. Figure 4.4 shows one of these models, constructed out of thick glass sheets. In 1903 Nathaniel P. Hill of Colorado and John R. Chamberlin of New York received patents in both the United States and Great Britain for a transparent mine model made of sheets of carved glass.[65] The pair noted in their patent application, "We are aware that attempts have been made to produce models or exhibits for this purpose, but we are not aware that any one [sic] has hitherto produced a model or exhibit in transparent material."

Hill and Chamberlin also filed for a patent on a machine for grinding the glass plates used in their models. The pair typically built models on a scale of one hundred feet to an inch, and used inch-thick glass plates to correspond with levels one hundred feet apart. Their models represented surface features as well as the underground workings. Modelers working on a Hill and Chamberlin model carved out the glass rather than painting or drawing on it as traditional sectional modelers typically did.[66] The primary audience for these expensive models, according to Chamberlin, was the executive associated with the mine; he reported that several occupied "a prominent place in the Director's room."[67]

A different form of sectional model eliminated the glass plates altogether, but retained the essential feature of creating a three-dimensional representation out of a series of two-dimensional slices. To address the optical problems of seeing through more than six or eight layers of clear glass, Swedish mining engineers created models for their deep multilevel mines that used a stack of wooden frames, strung with a thin wire grid, on which they attached copies of the workings in pasteboard. Each of these wire-strung frames, with its pasteboard mine workings, represented a horizontal slice of the mine in the same way that one pane of glass did in a glass-plate sectional model. The frames were then stacked, like the glass plates, to create the total model.[68] An earlier model in the British Museum of Practical Geology, made by T. B. Jordan, used a similar scheme of wires and workings, with each vein's excavations colored differently. "Although affording a perfect view of the general arrangement of a mine, this method is too cumbrous and expensive to be generally employed," noted Bennett Brough.[69]

Figure 4.4. This unusual sectional model used channels carved out of thick glass plates to represent the workings. The glass plates were then bolted together in sections. This technique for making models was patented in 1903, but never became widespread. From T. S. Harrison and H. C. Zulch, "Court Maps and Models," *Mines and Minerals* 29 (September 1908): 49.

Over time, enterprising model makers experimented most with the simple sectional model form, trying different configurations or construction techniques to make the models easier to build, more informative, or both.

Though most sectional models had a clear top to facilitate looking down through the layers of glass, occasionally enterprising model makers would place tiny models of shaft houses and surface works on the surface of the top sheet to represent the works on the ground.[70] This practice echoes, in three dimensions, the early mapping practice of drawing detailed miniature pictures of shaft and surface structures on vertical and longitudinal underground maps.[71]

Several model makers attempted to make the simple and inexpensive sectional glass model more three dimensional by adding material between the plates of glass. In 1909 mining engineer Edmund D. North described a sturdy glass sectional model he had made of the Montana-Tonopah mine. North used small brass clips to hold thin strips of glass vertically between the horizontal level plates. These vertical strips allowed North to portray stopes and intermediate workings as well as the regular levels, and had the added benefit of strengthening the glass model.[72]

Similarly, in 1912 mining engineer Maxwell C. Milton searched for a more satisfactory way to display operations at the Copper Queen Mine in Bisbee, Arizona. Like most model makers, Milton turned first to the traditional horizontal glass section model, due to the simplicity and cost-effectiveness of that design. Milton then cut and painted thick chunks of celluloid to represent the excavated portions between the levels represented by the glass plates. He placed these three-dimensional excavated stopes in position on the glass plates representing the levels, giving him a final product that combined design elements of both sectional and negative models.[73]

Some model makers, however, concluded that additional material between the plates was not worth the effort. Volney Averill built a sectional model made up of horizontal glass plates for the Tonopah Extension Mining Company that emphasized simplicity and low cost. In contrast to the practices of North and Milton, Averill deliberately avoided placing any three-dimensional workings between sectional plates. The engineer noted that such items made it more difficult to remove the plates to update them, and detracted from the "clearness" of the finished model.[74]

Well past the end of the period under consideration in this book, when it was deemed necessary to display the workings in three dimensions, glass plate models were usually the technology of choice. In 1969 Anaconda mining geologists decided it would be helpful to have a large model of Butte and the growing Berkeley Pit for the company office. The form they chose to execute was a glass plate model. Though it was built with twentieth-century innovations, such as aluminum rails for the glass and fluorescent tube lighting from the bottom, the basic form—mine workings drawn to scale on sheets of glass, placed at regular intervals in a rectangular frame—would have been immediately recognizable in Charles T. Healey's glass model of New Almaden, made nearly a century before.[75]

The sectional models described above used clear glass to make a series of two-dimensional drawings that collectively give the *illusion* of a third dimension. Mining engineers developed other types of models that represented mines in three true dimensions, but these lacked the ease of construction of the sectional glass plate models. *Negative models,* often called skeleton models or wire models, though less common than glass plate sectional models, were built by mining engineers to represent underground workings that were more difficult to show in two-dimensional slices. (We will analyze one such skeleton model from a 1914 lawsuit in the following chapter.) Negative or skeleton models were a mirror image of the underground, since they used solid materials to depict excavated space in the mine and represented solid mine rock as simply open space on the model. The models were usually constructed out of lightweight wood, copper wire, and occasionally cloth, wire mesh, or plaster-soaked strips.[76]

One difficulty with negative models was representing the mine workings accurately in a truly three-dimensional space. Model makers worked with the two-dimensional horizontal mine maps, vertical sections, and longitudinal projections, but in all cases work had to be careful and checked by technical experts for mistakes. Model makers also made trade-offs about how to best convert the features they wanted to display into three-dimensional form. For example, several of the models from the Butte District, made for the Anaconda Copper Mining Company for use in lawsuits and now on display at the World Museum of Mining in Butte, reveal that large stopes along steeply pitching veins were modeled by building up thicknesses of wood sawn to conform to the shape of the square sets in the vein as recorded in Anaconda stope records. The stacked block look of the stopes on these models gives the modern viewer an impression of a massive Lego construction. This technique might have reduced the aesthetic appeal of the model, but would have helped model makers complete their painstaking work more quickly.[77]

By contrast, other model makers felt free to impose greater artistic license on their models (perhaps supervised, or encouraged, by a geologist's projections). The wood model of the Anaconda-Neversweat mine, seen in figure 4.5, also from the Butte District and made for a lawsuit, depicts the complex vein structure and faulting in the Anaconda and Neversweat claims with beautiful, flowing carved sheets of wood, painted vivid red, blue, and yellow, to depict the veins and faults, with relatively minimal additional mine workings added in gray. Rather than the blocky look of some stope models, this model uses graceful construction to focus the viewer's attention on the geology at issue.[78]

Other model makers imposed three-dimensional grids on the spaces they were modeling to help them orient the miniature mine workings. These could be as simple as a set of numbered coordinates painted on the outer frame of a model, as seen on one complex skeleton model made for a Butte lawsuit in 1914.[79] Particularly in later years, when mine-wide coordinate systems became more common, these model grids might be adaptations of three-dimensional coordinate systems employed at the mine on its regular maps. While the three-dimensional coordinates were

Figure 4.5. This skeleton-style model was made for the *Colusa-Parrot* v. *Anaconda* lawsuit in Butte. Unusual for skeleton-style models, this depicts the vein as well as the workings. The original model is currently on display at the World Museum of Mining in Butte, Montana. From T. S. Harrison and H. C. Zulch, "Court Maps and Models," *Mines and Minerals* 29 (September 1908): 53.

often projections, some model makers actually constructed the coordinate grids out of wires, creating a series of squares or cubes into which modelers could place the models of the workings.[80]

For example, the United Verde Mine, at Jerome, Arizona, created a skeleton model in 1910 based on the coordinate mapping system in use throughout the mine. United Verde engineers manufactured elliptical steel frames and strung them with piano wire representing the coordinate grid, weaving the wires together like the strings on a tennis racket. They then placed strung elliptical frames horizontally, one atop the next like the glass sheets in a sectional model, on a set of vertical rods with spacers, so that the frames held the coordinate wires in their proper three-dimensional orientation. Drifts and workings were traced from the mine's maps on to pine boards, then cut out and glued in place on the wire grid. Small notches were made in the wooden workings where necessary to allow the coordinate wires to pass, and the workings were painted once set in place. The top level represented the surface in lifelike detail, and was constructed from built-up wooden contours constructed in the fashion described by Lesley decades before. (The model maker sprinkled ground-up rock over wet glue to depict the surface geology as it appeared on the ground.) They built the model on a custom table that contained a bulb behind ground glass embedded in the surface to light the model from below.[81]

Figure 4.6 shows a skeleton model constructed circa 1917 for an Idaho mining lawsuit. In this model, the top represented the surface, depicted with contour lines as on a topographic map, complete with painted elevation numbers. The bottom of the model had a two-dimensional plat map of the boundaries of the mining claims at issue in the lawsuit, along with a title block to identify it. The mining claims were also outlined on the surface topography in wire, so that the borders would be visible when the viewer looked down on the model from above.[82] The tunnels were mostly made from wire, and the stopes, or excavated areas, appear to have been thin pieces of cardboard or wood. Each feature potentially of interest in the lawsuit was labeled with a paper tag. Though it is impossible to be certain when viewing the black and white photographs taken of the model, a legend on the bottom map suggests that the different tunnels and stopes were probably colored according to their elevation. This would follow the ordinary practice employed on composite two-dimensional mining maps of giving each level a different color as a way to tell them apart.[83] This

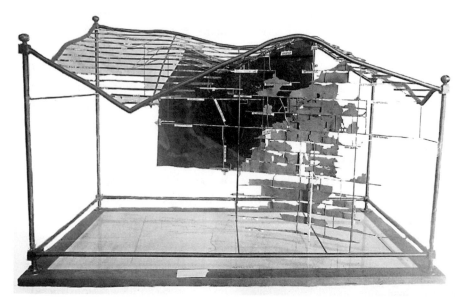

Figure 4.6. This skeleton model, made for the same lawsuit as the glass plate model in figure 4.3, is a classic example of depicting the excavated workings in solid materials, and leaving unexplored rock as open space on the model. Note the heavy frame needed to protect the fragile model. "Plaintiff's Exhibit 36," Bunker Hill / Pintlar Corporation Collection, Manuscript Group 413, Box 337, folder 5933, Special Collections and Archives, University of Idaho Library, Moscow, ID.

model's heavy outer frame was this model maker's way of solving one of the primary problems with skeleton models, which was that they tended to be extremely fragile. Even with the unusually heavy frame to stabilize it, additional wires connected the workings to the frame for added strength. What appears to be a curtain or partial background is not part of the model. Careful examination of the full series of five photographs (part of the same collection as figure 4.6) taken of this model shows that the photographer attempted to hold a cloth backdrop against the model to exclude the background scene. The background must not have provided adequate contrast for the photographer's taste, however: the negative was carefully filled in with opaque ink, since dark ink on the negative yielded pure white on the print, to blot out the background. (A tell-tale touch of sloppiness with the pen on the negative can be seen on the structural members when we view the photo in figure 4.6 under magnification.)

Negative or skeleton models were certainly more difficult to construct and update, which meant that they were most commonly used in a legal context. However, negative models were occasionally constructed to be used by the mining company for some of the same purposes for which a glass model might be used, if a skeleton model was most useful for depicting the layout of the mine and some compromises were made in modeling. In 1910 the North Star Mine in Grass Valley, California, used a negative model made of white-painted copper wire to help portray the mine as a whole and plan broadly for new development work. In many other mines, if the company were to decide to use a model at all for such a purpose, it would likely have been a glass sectional model, made directly from the working maps. In the North Star, however, the vein and workings were flatter than in many mines. To represent them on a glass model, the glass would have to be large and lots of it would be empty. Instead, mining engineer Arthur D. Foote worked out a model made of copper wire, bent to represent the workings, inserted into wooden vertical boards representing the coordinate planes used by the mine's maps. Engineers could easily remove and update or replace each piece of wire. Foote bypassed one of the main challenges of negative models by choosing to almost completely ignore the stopes and instead focus on portraying the levels only.[84] Foote, the mining engineer who came up with the idea and construction plan for the North Star model, was by that time near the end of his career—one that seemingly featured model-making throughout, with service at New Almaden and a glass model in 1880s Leadville, as mentioned above.

Wires that composed part of negative models could also depict usable information. For example, around 1912 W. G. Zulch constructed a small model of a tiny portion of a mine in Cripple Creek for use in a lawsuit between the company and one of the teams that leased ground within the mine. The leasers possessed a lease on one vein, but when that vein intersected other veins, the leasers began mining the wrong one. Zulch's model showed only the portion of the two stopes and three levels that were involved in the lawsuit. At a scale of five feet to one inch, the model used horizontal wires as contours to show the precise size and shape of the stopes as they had been incorrectly worked by the leasers. Zulch reported recording the information used to create the model directly from the stopes themselves.[85]

Though wood and wire were the most common materials used to create negative models, other materials were used as well. For example, a model the Della S. Consolidated Mining Company, of Aspen, Colorado, made to defend itself against an apex lawsuit from the neighboring Standard Mine was a skeleton-style model with an interesting twist. The surface of the ground, at the top of the model, was, in this case, represented by a sheet of plate glass that was molded to a wooden form depicting the surface topography. This model also had a back board that showed a vertical section along the vein, out of which the three-dimensional workings of the model were supposed to extend.[86]

In some cases, models made use of different materials in order to show both the excavated mine workings as well as the veins from which those workings were developed. One of the models made in a 1901 lawsuit between two mines in Grass Valley, California—the Grass Valley Exploration Co., and the Pennsylvania Mining Co.—used a series of thin strings to represent later veins, and reserved the traditional wood for workings.[87] Similarly, an undated negative model that may depict the Butte & Superior Mining Company's Black Rock claim, probably dating from circa 1900, uses wide strings or ribbons to show the outer boundaries of the steeply dipping veins, which permitted the modeler to use carefully carved wood to depict the workings *within* the veins. This model also uses a curtain of thin wire, stretched tight, to project the vertical line of a claim boundary down into the workings below.[88]

Mining engineers and geologists occasionally developed new ways to represent two-dimensional data in three dimensions, particularly as they generated greater amounts of information after the turn of the twentieth century. The opening of the so-called porphyry copper deposits (very-low-grade ore spread over a wide area, mined, often using open pits and steam shovels) was possible in part because mining companies used diamond drills to sample large orebodies inexpensively. The use of diamond drills to prospect for orebodies made a significant difference in the ability of engineers to plan future actions on the basis of known reserves, but the amount of data generated in the course of a drilling campaign could be difficult to visualize. One solution was to make a model of wooden dowels, to represent drill holes, set in pegboard.[89] A 1910 article in the *Engineering and Mining Journal* suggested that dowels in such a model should have one color to indicate the drilling progress, and another to indicate ore found in the drill holes.[90] The Live Oak copper mine in Miami, Arizona, used a different solution. The modeler in this case used glass tubes, in the same fashion as dowels in pegboard, but connected the painted parts of each vertical tube to its neighbors with horizontal strips of lightweight bamboo. By connecting the tops and bottoms of the exposures of ore in each drill hole, the bamboo outlined the orebody in three dimensions, just as the drills discovered it.[91]

While most dowel models represented the results of drilling beneath the surface, modelers could use such models in other types of mining problems where they could array two-dimensional information at a known point spatially to provide a three-dimensional picture. A model of this type produced by the C&H in the 1910s is shown in figure 4.7.

Figure 4.7. Painted dowels set in pegboard were the basis of this model of stamp sands in Torch Lake made by Calumet and Hecla. The sands had been dumped as tailings into the lake in the nineteenth century, but in the early twentieth century the availability of better recovery methods prompted the company to dredge them up for reprocessing. From the Koepel Collection, Keweenaw National Historical Park Archives, Calumet, MI.

Since the start of milling operations around the time of the Civil War, the C&H had been discharging copper sand tailings from its mills into the deep water of Torch Lake. These tailings still contained native copper in particles too small to grind and recover. As milling technology advanced in the late nineteenth and early twentieth centuries, the C&H became hopeful of tapping the tailings as a new source of copper, using new techniques to recover metal that old technology only left behind.[92] Part of the trouble was identifying the size and value of the tailings piles deep beneath Torch Lake to see whether the necessarily extensive investments in new technology needed to tap the deposits would be justified.

The C&H engineers constructed an innovative model that drew on diamond drill hole models to help them understand and direct the dredging problem. As seen in figure 4.7, the model consisted of a flat horizontal pegboard with painted vertical dowels placed in the board at regular intervals. The model itself was large, occupying the top of a sturdy map cabinet in the engineering office, and representing more than 150 acres of underwater terrain. The company used data from a United States coastal survey conducted before the mills started operating to establish the bottom contours of Torch Lake. Next, C&H took standard nautical soundings every fifty feet in a grid pattern, aided by the establishment of survey points on the shore and floating buoys at specified locations. C&H engineers then plotted these data on the pegboard. Each dowel represented a sounding location, and was cut to indicate the current height of the underwater surface above the pegboard's control baseline. Modelers than painted the dowel white from the pegboard base to the depth of the original lake bottom, as recorded in the government survey. The remaining portion of the dowel, from the original lake bed to the present lake bottom, represented the tailings; modelers painted it a dark color. The position of the lake level and the control buoys were represented by tall white dowels with small labels attached to the top. The company conducted underwater soundings frequently once the dredging operation commenced, and updated the model with the new information. Engineers used the model to create cross-sections that were used by the dredge operators so they would know how to proceed, and also kept records of the gradual diminution of the tailings in two-dimensional cross sections made from the model.[93]

The prospect of additional copper profits sitting at the bottom of Torch Lake, as brought vividly to life by the dowel model made by C&H

engineers, was of tremendous importance to extending the life of the mining company. Even before the recovery operations commenced, the possibility of profits from the tailings became a serious issue in a series of lawsuits in 1911 that sought to prevent the C&H from consolidating its control over several neighboring firms. Stockholders of the other firms refused to believe C&H's contention, brought to life on the dowel model, that the rich underwater tailings should be counted as a recoverable asset, and thus should increase the value of C&H stock in the merger.[94] The 1911 merger ultimately failed, but in fact the C&H had actually underestimated the value of the tailings. From the beginning of the dredging operation in 1915 until the sands were exhausted in 1949, the C&H recovered and treated more than 37 million tons of copper-bearing tailings, yielding more than 400 million pounds of copper, and a total profit of more than $31 million.[95] Such profits were the result of an expensive and innovative dredging and milling system, but one glance at the photograph (figure 4.7) is sufficient to understand the powerful role played in this operation by the aggregated information displayed in a three-dimensional form on the dowel model.[96]

Occasionally modelers combined features of different types of models. Three models of salt mines in present-day Austria, probably made in the late 1850s, were on display at the London Museum of Practical Geology by 1865. Bergmeister Ramsauer, the mining engineer in charge of the works, made the models at a scale of four hundred feet to an inch. These models combined features of block models and sectional glass plate models. On first viewing the models looked like the ordinary block type, with the surface represented accurately on the top and geologic sections painted on the vertical sides of the model. However, when the top of the model was removed, the user looked down into a series of horizontal glass plates with certain features of the workings colored in. Thus the Ramsauer models combined both the block model and the glass plate modeling traditions.[97]

In contrast to underground maps, mine models were most useful in helping non-miners visualize complex underground geology. In his 1855 book John R. Leifchild spent several pages attempting to describe the complex underground geology of Wheal Peever, a Cornish tin mine. Finally, perhaps sensing that his explanation and two-dimensional geometrical drawings were inadequate, Leifchild noted, "Such complicated phenomena are only to be clearly apprehended by a model. I fear the reader may think I am here presenting to him the *pons asinorum* of the mineral Euclid."[98]

After the turn of the century, authors of treatises on mining engineering advocated models for their utility in helping visualize the underground, especially for so-called practical miners (i.e., those potentially without the benefit of a university education in mining engineering). William H. Storms, in a book aimed at this "practical" audience, interrupted his 1909 discussion of the block-caving method pioneered by the Homestake Mine in South Dakota to offer his opinion as to the value of three-dimensional models of the underground: "A good model of the workings of a mine is of great value in laying out work and in studying mine methods," he wrote. Storms believed that such models were of particular help to superintendents, foremen, and managers generally, in part because models "make it possible to present the broader problems involved within a space immediately under the eye." The author pointed to a five-foot-square model of an underground stope used by the Homestake to help them develop new mining methods, and he also mentioned a plaster of paris model, sawed into sections, created by the Treadwell Mine in Alaska. Storms concluded his side note by comparing models with the more familiar mine maps, while keeping his practical audience in mind: "Maps, of course, have similar advantages, but a properly constructed model is better for the purpose of studying practical mining problems than the best map."[99] Storms' advocacy of models over maps for his practical audience may have been in response to the increasingly fragmented and specialized nature of underground maps. These larger-scale maps showed less ground, and thus were more difficult, especially for those not formally trained as mining engineers, to reconstruct in their mind's eye as a three-dimensional whole. In such a situation, a three-dimensional model, that necessarily was smaller scale, could provide the overall view that maps increasingly could not, but that understanding mines as increasingly large and interconnected systems seemed to demand.

Models also could be useful to address audiences familiar with mining but not trained to read mine maps, such as non-Americans or ordinary miners. Mining engineer C. L. Severy created a glass model of the Poderosa Mine in Chile to help his South American shift bosses and foremen understand the workings as a whole. "Most natives of South America cannot read a map, no matter how well prepared," groused Severy, but a mine model gave his mine foremen a usable "perspective" on the mine and its orebodies in all their complex glory. Studying the model also helped the locals eventually learn to read a regular mine map.[100]

Models could also be used to help miners who could not read mine maps. For example, in the later life of the mines of the Tonopah Mining Company in the 1930s and 1940s, after the boom was long over, leasers did the mining. To facilitate their work, the company maintained an up-to-date glass plate model with the workings and the geology of the mine in the mine office. Before agreeing to lease a section of the mine, the leasers could come into the office and use the model to help them pick a potentially favorable spot. Leasers tended to be practical miners, without any engineering education, and they likely would not have been able to derive the same use out of traditional mine maps.[101]

With a case relating to mining law at hand, the courtroom—where non-technical judges and juries applied vague laws to specific underground conditions—became one of the most important sites for the visual work at which models excelled. "In addition to the practical value and advantage of mine models to the geologist and miner, such models will frequently be found of great advantage in suits at law, in settling mining claims and damages in dispute, when ordinary plans are unavailing," advised Bennett Brough in his classic work on mine surveying.[102] In a legal context, the visual power of models outweighed their drawbacks of expense, small scale, and inflexibility. Chapter 5 closely examines one particular mining trial and its models in its legal and geological context. Accordingly, the treatment here of the use of mine models in litigation will be brief, but it is worth recalling that legal struggles formed the primary context in which underground models were made and used during this period.

Mining engineer and geologist David W. Brunton, whose efforts helped establish the geological mapping at Anaconda were recounted in chapter 3, was also well respected in the courtroom as an expert witness. Brunton succinctly argued for the importance of visual representations in mining trials: "The best method of placing actual mine conditions before a judge or jury is by some graphic method of visualization. Verbal descriptions of mine-workings convey little or nothing to a man who has never been underground. . . . [T]heir sympathies, [are] always with the side that they understand best; hence the necessity in a mining suit for introducing models, colored maps, and anything that will enable the jury to visualize conditions better than they can from verbal descriptions."[103]

The early glass plate models made for the *Eureka-Richmond* suit in 1877 were famous, in part because of the attention garnered by the suit, but also because of the size and complexity of the models. From the 1890s it was rare that a mining lawsuit of any importance proceeded to trial without competing models. In the early twentieth century, there are examples of model building as a part of a saber-rattling strategy by one mining company to convince another to settle their disputes, rather than go to trial, and modelers often made models for pretrial hearings as well.[104]

Most models used in mining litigation were technical models, such as glass plate models, skeleton models, or block models, but on at least one occasion museum model-making practice found an application in a mining lawsuit. Sometime near the turn of the twentieth century, Brunton was involved as a consulting engineer for the Golden Cycle Mine in an apex suit with the Vindicator Mine, both of which were located in the Cripple Creek gold mining district of Colorado. He anticipated that the presence of the vein in the discovery shaft would be an important point in the case (since, in the apex law, the discovery shaft must actually find a vein). Brunton professed to have been stuck on the problem of conveying the appearance of this vein in the shaft to the jury. The construction of realistic rocky backgrounds for animal tableaux at the Denver Natural History museum by a New York museum modeling expert inspired Brunton's solution. The mining engineer hired the model maker, after he had completed his work for the Denver museum, to make a detailed model of the vein in the shaft. The vein model, "an absolute reproduction of the discovery shaft in form, texture, and color," according to Brunton, took two months to complete. The two companies settled in order to avoid a trial, but the opposing lawyers were impressed by the model, according to Brunton. "The gentlemen on the opposite side of the case, when once the suit was settled, were quick to acknowledge that this reproduction of the shaft would have meant their certain defeat had the case come into court." Brunton enthusiastically recommended the future use of such models, but it is unknown if other mining litigation experts took up Brunton's suggestion.[105]

By the end of the period covered in this book, as mining geologists assumed decisionmaking power at some of the largest mining firms, underground models occasionally began to be used in actual mining operations. In 1903 the Alaska-Treadwell Mine gained attention by creating a sectional model out of wood for use in daily operations. The model was the same scale as the company's mine maps, and monthly carving work kept the work up to date. "The model is studied carefully by the foreman and shift bosses, and it is said to have proven of great value in the operation of the mine."[106] The *Mining and Scientific Press* noted the novelty of the company's use of such a model for a purpose outside litigation: "[P]ractical men will approve such an innovation."[107] By this time, however, the term *practical miner* had come to mean one who had not been formally educated as a mining engineer, such as a foreman or shift boss. In other words, the Alaska-Treadwell model would be useful to speak to an audience that might not be able to make effective use of mine maps.

A 1906 description of the system used by the Utah Consolidated Mining Company to work their underground copper mine at Bingham Canyon noted that, in addition to several sets of different underground maps that were meticulously maintained, the company also had a glass plate model showing both the works and the orebodies, "which is also brought up to date from time to time."[108]

Mining engineer Maxwell C. Milton, of the Copper Queen in Bisbee, noted in 1912 that mine models were "becoming more and more a part of the working apparatus of large mines." The Copper Queen experienced a slow-burning underground fire in 1911 in the support timbers buried under waste rock in an abandoned working chamber in a portion of a large, irregularly shaped orebody. The unusual shape of the ore, however, made it difficult to determine how and where to cut off the fire from the unmined portions of the mine. To help them solve their problem, the Copper Queen engineers created a sectional model of the area (from celluloid sheets). This model helped the miners attack the issue, and the usefulness of this small model convinced them the Copper Queen engineers to make a large model of the entire mine.[109]

In conclusion, models—like maps—were critical to the development of the visual culture of mining. Without the ability to see the inner workings of the unexplored areas of a mine, models served a valuable purpose for those looking to understand the complex systems beneath the earth. Models were quite complex, and required relatively large amounts of time and money to construct. Because of their inflexibility, and expense, as well as the expertise required to create models, these tools were used most extensively in courtrooms and classrooms, where their ability to portray the invisible spaces of the underground to audiences without a technical background was of primary importance. In chapter 5, we will examine the role played by a series of dueling models in a large mining lawsuit. In chapter 6, our focus shifts to the role that mine models played in educating young mining engineers at schools of mining, and in educating the public at expositions and museums.

These powerful artifacts—technical models—sometimes possessed a kind of strange beauty. "Models hold promises—the delight, for example, of seeing inside, and of taking things to bits. . . . The ability to contemplate normally-hidden structures . . . has its own frisson," noted historian Ludmilla Jordanova.[110] Certainly the artifacts could carry those sorts of associations for the contemporaries of their makers. Marjorie Brown was a twenty-year resident of Tonopah, and was married to Hugh Brown, a very well-known mining attorney and a member of the legal team in the *Jim Butler* case. She chose a glass model as a metaphor to sum up her life in Tonopah:

> Tonopah geologists made models of their mines. On thin glass slides, some of which hung vertically in slender grooves while others lay horizontally on tiny cleats, all the workings of the mine were traced to scale with colored inks. When you stood in front of the model and looked into its serried sections, you seemed to be looking into the earth with a magic eye. Here the shaft dropped down from level to level through ore and country rock; here were "drifts" and "stopes" and "crosscuts" with every foot of ore blocked out; and here you traced the meandering vein, noted where it petered out or widened into richness unimagined as it continued into regions still unexplored.
>
> They were beautiful things, these glass models, made by skilled craftsmen, often works of art. When I think of Tonopah, the memory is like such a model of a life I was privileged to live, unique and gone. . . . [My memories of Tonopah], so full of meaning for me, were a vein of ore whose richness increased with depth.[111]

chapter 5

Models and the Legal Landscape of Underground Mining

Visitors to the W. M. Keck Geology Museum, housed in the Mackay School of Mines building on the campus of the University of Nevada–Reno, are confronted with a bewildering array of interesting stuff. Long cases full of mineral specimens, both common and rare, form the heart of the museum's collection, but the second-floor balcony holds treasure of a different sort. In a corner, there's a mining cage—a precarious-looking phone-booth-sized open platform used to whisk miners and material to the surface from thousands of feet underground. Nearby rests a bit of ancient mine lumber, twisted and squeezed by the weight of overhanging rock into a barely recognizable version of its original form. And along one wall, secure under a Plexiglas cube, is a riotous jumble of wood, wire, and paint, its combination of amorphous shapes and geometric lines suggestive of some bizarre accident involving an erector set and a bucket of candle wax.

This is a mine model, built in 1914 for a mining lawsuit (figure 5.1). It depicts, to scale and in three dimensions, the underground workings of a portion of the Tonopah, Nevada, mining district. It was built by the defendant in a complex lawsuit over who had the right to mine more than $500,000 worth of silver ore. Thanks to good surviving documentation, this model, and the trial in which it was used, can give us an unusually clear window into the use of three-dimensional visual representations in mining lawsuits.[1] In this case each side utilized models, but it is clear from the historical record that the competing models were used in far different

Figure 5.1. This skeleton model was created by model maker George Pierce for the West End Consolidated Mining Company in 1914, for use in the firm's lawsuit against the Jim Butler Tonopah Mining Company. After the trial and appeals were complete, it was donated by the company to the museum of the Mackay School of Mines at the University of Nevada–Reno, where it remains on display today. Photo by Eric Nystrom, 2012.

ways by the litigants. Here, models helped create and reinforce competing interpretations of a specific underground landscape. An expensive lawsuit and ownership of a silver-rich vein hinged on how the model-mediated particular details of that underground space did or did not fit with the established parameters of American mining law.

To a modern observer, it is difficult to imagine this artifact as useful for much of anything, yet the model played a critical role in structuring the final outcome of the case. The verdict resulted in prosperity for one company and ruin for the other; the case also established a mining law precedent that the US Supreme Court affirmed.[2]

Since all mining law was interpreted in light of local geological conditions, it is important to set the stage for the discussion of the case with a brief account of the history of mining at Tonopah. In 1900 prospector

Jim Butler discovered rich silver ore at the future site of Tonopah, on the flank of a mountain in central Nevada. The following year he returned with friends and began to develop his claims in earnest. The potential of Tonopah brought a huge crowd to the new camp, and also brought the attention of Eastern capital. Butler and his partners sold the original claims to the Tonopah Mining Company, backed by Philadelphians, and the company commenced mining in January 1902. The Tonopah Mining Company held some of the best claims, but many others had been staked by prospectors attracted by Butler's initial discoveries, some of which turned into profitable mines as well. The boom period lasted through about 1905; after that, Tonopah mining became less speculative and more business-like. The district's period of peak productivity was about 1908–1924. The very best years occurred almost in the middle of that period—total output from the mines exceeded 500,000 tons annually in 1913, 1914, 1915, and 1918. During these years of peak productivity, there were over half a dozen major mining companies, and many more small ones. Some of them were interrelated, sharing board members, managers, or even mining personnel, but no one company completely dominated Tonopah as was common in other districts. Figure 5.2 shows the ground controlled by the various companies as of 1915, which even extended directly under the town itself.[3]

One of these businesses was the Jim Butler Tonopah Mining Company, named after the discoverer of the original deposits. It was formed in 1903 to work sixteen claims that were located southeast of the original discoveries (see figure 5.2). Jim Butler himself was president of the namesake company until 1911, but throughout this initial period the Jim Butler was operated by the Tonopah Belmont operation, which was controlled by many of the same Philadelphians as those who bought out Butler's original discoveries. From its formation in 1903 until 1910, the Butler was mined in a relatively limited fashion, with low output. In 1910 miners found more-valuable deposits underground, and the Butler turned a profit for the first time in 1912.[4]

The West End Consolidated Mining Company was based on the West End claim, which was located in 1901 by several of the original leasers of Jim Butler's initial discoveries (prior to the beginnings of Philadelphia control). They formed a locally owned company, which was uncommon in the area, but a lack of capital restricted production. In 1906 the miners encountered high-value ore—averaging $62 per ton—that justified

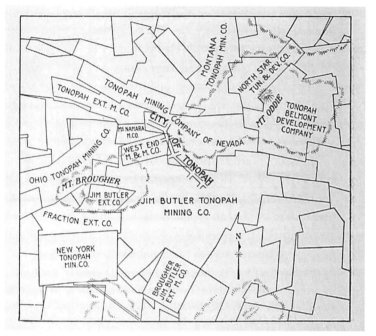

Figure 5.2. Map of mining companies in Tonopah at the time of the lawsuit. The West End is a small claim in the center left, and the portions of the Jim Butler ground in dispute are located just to the south. Note the jagged end line of the West End claim. From "Apex Litigation at Tonopah," *Engineering and Mining Journal* 99 (1915): 660.

incorporation and a public stock offering to raise working capital. Francis M. "Borax" Smith bought most of the shares and became president of the company, but the original owners were still involved with active management of the mine.[5] The West End settled a dispute over extralateral rights with the MacNamara, the claim immediately to its north, in 1908. Figure 5.2 shows the MacNamara north of the West End claim, which borders the Jim Butler holdings to the south, all directly under the main streets of Tonopah. By 1914 the West End had earned a solid reputation. The mine shipped ore steadily, and management was praised for making decisions based on conservative estimates and pushing a forward-looking development strategy.[6]

These two neighboring businesses, the Jim Butler Tonopah Mining Company and the West End Consolidated Mining Company, were the antagonists in a high-stakes 1914 lawsuit. The fight began in February

1914, when miners employed by the Jim Butler broke through into work-ings made by the West End in Jim Butler ground. It turned out that the West End had mined more than 55,000 tons of ore beyond its side bound-ary in Butler territory.[7] After negotiations to prevent litigation and to equally divide the ground failed, the two companies prepared to battle it out in court.[8]

The legal basis for the dispute was a branch of mining law known as extralateral rights, which under certain conditions permitted miners to follow a vein from their own claim underneath a neighboring claim as the vein dipped into the earth.[9] Extralateral rights were based on the idea that the locator of a claim on a lode could follow that vein underground, through the side lines, even if it went into adjacent properties, as long as the top or "apex" of the vein was on the locator's claim.[10] The exact origins of apex rights are obscure, but precedents existed in medieval England and Germany.[11] The apex right, as it evolved in the American West, permitted the miner to mine deeper along the *dip* of the vein, out the side lines of the claim, but limited mining along the *strike* of the vein to imaginary vertical planes projected downward from the end lines.[12] The idea of the extra-lateral right was to protect the prospector—no X-ray vision could predict where a vein might go once it departed the surface, but the extralateral right attempted to ensure that the diligent prospector would not lose title to his mine simply because of an unforeseen geological twist. These prin-ciples were implicit in the Mining Law of 1866, and were explicitly written into the Mining Law of 1872. In order to possess this extralateral right, the claim had to include the apex of the vein within its surface boundaries, and also meet all of the other requirements (size, fees, shape, and so on) spelled out in the law.[13]

A mining company needed an apex in its claim in order to have extra-lateral rights, but what exactly was an apex? The statute did not define the word, and it was not a traditional mining term, so no historical meaning could serve as a guide. Congress apparently had an idealized type of fis-sure vein in mind in using the term, and in such a case it might have been clear enough what the apex of the vein was, but the geology of mines was rarely so simple. As a result, the courts gradually refined the term *apex* as they decided on lawsuits over extralateral rights where existing precedent was unclear.

The West End believed they had an apex inside their claim that gave

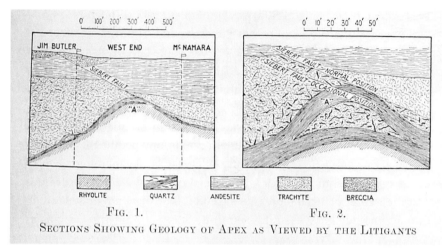

Fig. 1. Fig. 2.

SECTIONS SHOWING GEOLOGY OF APEX AS VIEWED BY THE LITIGANTS

Figure 5.3. A vertical cross-section of the vein at stake in the trial, showing the anticlinal shape and the Siebert fault. Compare this with the side view of the West End model in figure 5.1. From "Apex Litigation at Tonopah," *Engineering and Mining Journal* 99 (1915): 661.

them the right to follow their vein into the Jim Butler, but the vein did not look like Congress' ideal case, slashing through rock at a steep angle until it reached the surface. The West End's vein had a shape more like a handkerchief that someone pinched in the middle to pick up off a flat table. Figure 5.3 shows one vision of the cross-section of the vein. Geologically speaking, this inverted-U shape is known as an anticline.

The problem for the West End was that earlier courts had ruled, in a series of decisions known as the Leadville cases, that anticlines were not enough to constitute apexes. In the Leadville cases, the courts concluded that a blanket vein that merely rolled or undulated had no true apex, and therefore no extralateral rights. A "terminal edge" was needed to have an apex.[14] The West End attempted to dodge this precedent by arguing that, despite appearances, their vein was not an anticline, but was instead two *separate* veins that happened to come together at the top—two veins, two apexes (in the same place), and thus extralateral rights in both directions. During the trial the West End's lawyers and experts were careful to always use language that reinforced this interpretation. They never spoke of the whole structure as a vein; instead, they talked about the "North dipping vein" and the "South dipping vein"; likewise, the peak was always a "junction" or the apexes, never "the anticline." The Jim Butler countered that since the vein material was continuous from one branch over the top and down the other side, the vein was clearly an anticline, and thus, by virtue of the precedent established in the Leadville cases, should not possess either an apex or extralateral rights. The Butler team spoke of "the vein" as a whole and frequently referred to the "anticline," in a mirror of the West End's effort to use language that supported their interpretation of the law. Both sides knew the precedent of the Leadville cases would be significant, though the Leadville cases were based on veins that were very different from those in the Tonopah District. The West End hoped that the difference was significant enough to prevent the Leadville precedent from applying to the Tonopah situation, and instead to make its two-vein theory supportable. By contrast, the Butler believed that the situation was not so different as to invalidate the general point of the Leadville cases.

The stakes were high. All told, the ore in dispute was valued at about $500,00, and on top of that, Nevada law allowed for triple damages in cases of mining trespass. The lawsuit seemed to promise prosperity if won, ruin if lost; the future seemed to hang in the balance—so both companies set out to assemble the best legal talent money could buy.

The two mining companies assembled a "grand galaxy" of lawyers and experts to defend their claims.[15] Each side's team consisted of four different types of experts. Directing the overall strategy were hired attorneys with specific expertise in mining law and extralateral rights. These lawyers were assisted by the normal attorneys for the mining firms, who generally had less-specific expertise in mining law but who were more familiar with the internal operations of the company and the local context. Both sides hired eminent expert witnesses, generally with national reputations, to testify about geology and engineering practice, and thereby connect the specific details of the case with broader scientific theory. The teams were rounded out by local engineers, whose value in court was their intimate knowledge of the specific mines under discussion. Together, these four types of experts attempted to present a coordinated vision where legal arguments, visual representations, and geological facts worked together to make their interpretation the most compelling to the judge.

In the Jim Butler–West End trial, the expert mining attorneys were an especially distinguished group. The Jim Butler retained the biggest star of all, "Judge" Curtis H. Lindley, author of the most famous and important treatise on American mining law to that time.[16] In fact, the third (ultimately final) edition of Lindley's text was published in January 1914, less than a year before the beginning of the trial.[17] Lindley's father, also a lawyer, moved to California during the Gold Rush in 1849. Lindley was born in Marysville, California, in 1850, which was at the time one of the major centers of mining activity on the Mother Lode. Many of his teenage years were spent on the Comstock Lode in Nevada, and he later served as a hoisting engineer before he studied law and began practicing in California. He briefly served as a magistrate in Amador County, earning him the lifetime sobriquet "Judge." When he was not elected for a second term he turned more specifically to the study of mining law. Relatively few books were published on mining law at the time Lindley began his work, and "the few works that had appeared were little more than digests of the statutes and the few cases the courts had then decided. They could hardly be dignified with the title of treatises."[18] Lindley first published his monumental work *A Treatise on the American Law Relating to Mines and Mineral Lands Within the Public Land States and Territories and Governing the Acquisition and Enjoyment of Mining Rights in the Lands of the Public Domain*, commonly known simply as *Lindley on Mines*, in 1897. Later heavily revised editions appeared in 1903 and 1914. Lindley was apparently "such a stickler for the proprieties that he would never quote from or refer to his own book, and never allowed it to be brought into the court-room, even by his associates, when he was present."[19] Even so, the text became the widely acknowledged authority on American mining law; US Supreme Court justices even quoted it extensively in relevant opinions. The third edition of *Lindley on Mines* (1914) was considered the best of them all, a monumental work to cap a distinguished career. Horace V. Winchell, the famous mining geologist we met in chapter 3, who testified for the West End as a geology expert opposite Lindley, termed the book "lucid," "unambiguous," and "indispensable" in a review published in September 1914.[20] Eminent mining engineer Rossiter W. Raymond, himself an expert on mining law, heaped even more praise on Lindley's "tru[e] magnum opus." In the most widely read journal for mining engineers, Raymond described the third edition of *Lindley on Mines* as a "magnificent treatise," written with "candor, lucidity, and forceful suggestiveness," a work that exuded "comprehensive and classic excellence."[21] Lindley was also heavily involved in professional and civic causes. He was active in the San Francisco Bar Association, helped organize the California State Bar, and served as its first president. A political progressive, Lindley was one of the leaders of a reform movement in San Francisco, played a strong role as lead counsel in the effort to dam the Hetch Hetchy Valley to create a water supply for the city, was a director of the Panama-Pacific Exposition of 1915, and served as Park Commissioner for San Francisco.[22] Lindley was a friend of Herbert Hoover, who met the lawyer when he was engaged to help Lindley prepare for a lawsuit in the Grass Valley (California) District in 1896–1897.[23] Later, when Hoover was appointed head of the Food Administration during World War I (Lindley had facilitated the appointment in part, by introducing him to friends in Washington), the mining engineer persuaded Lindley to come to Washington to take charge of the legal work of the department. And Lindley was a poet, of sorts: he wrote doggerel verse during the Jim Butler–West End trial, which was printed by the local newspaper.[24]

Second in command of the Butler's legal strategy was William E. Colby, Lindley's trusted assistant and a recognized authority on mining law in his own right, who had long and successful experience with apex litigation. Colby began working for Lindley in 1907, and took over the older lawyer's practice when Lindley died in 1920. Colby was most famous, later in life, as a longtime officer of the Sierra Club. Colby was the lead counsel for the anti–Hetch Hetchy Dam movement, even though he worked out of the same offices as dam advocate Lindley. (Lindley hoped to keep that quiet, to avoid any sense of impropriety of having both sides represented by the same firm.)[25] Colby certainly understood extralateral rights well—his 1916 four-part law review article on the apex issue, produced after the initial trial in Tonopah but before the *Jim Butler* v. *West End* case was finally settled in the US Supreme Court, was, according to a modern legal scholar, "perhaps the most articulate and certainly the most comprehensive defense of the apex law."[26]

Mining lawyer William H. Dickson, of Salt Lake City, headed the West End Consolidated's legal team. A native of New Brunswick, Canada Dickson spent eight years as a lawyer on the Comstock in Nevada. In 1882 he moved to Salt Lake City, Utah Territory, and was appointed US attorney in 1884. As part of the Gentile minority who occupied a majority of the federal territorial offices, Dickson zealously prosecuted Mormon polygamists under the federal Edmunds Act.[27] He so enraged members of the Mormon community that, in 1884, someone lobbed glass jars of human waste through Dickson's window, which broke on the walls and carpet. In 1886 the son of one of the Mormon leaders struck Dickson in the face during a personal meeting.[28] In 1887 Dickson retired from the US attorney post, citing the low salary, and resumed a successful (and lucrative) private law career. Over the next decades Dickson developed an excellent reputation for mining law, especially apex suits, and tried many famous cases.[29] Dickson never gained Lindley's national notoriety, probably because he never published a treatise, but he seems to have had a positive reputation as a mining lawyer. Like Lindley, Dickson had less than a decade left to live, but continued to work until the end.[30] Dickson was familiar with Tonopah because of his business interests in the camp. He served as a member of the board of directors (along with A. C. Ellis) for the Montana-Tonopah Mining Company, and briefly held an interest in some of the earliest claims located in the camp.[31] Dickson later unsuccessfully defended the Tonopah company that was formed to work his claims in a series of boundary lawsuits. Ironically, at least two of those claims, the Stone Cabin and the Wandering Boy, were later absorbed by the Jim Butler Tonopah Mining Company.[32] In short, Dickson had a long history with Nevada, Tonopah, and mining law.

Working closely with Dickson on the West End Consolidated strategy was mining lawyer A. C. Ellis Jr., the son of an early Nevada lawyer and politician. Ellis moved to Salt Lake City in 1892, and entered into a multidecade legal partnership with Dickson and the elder A. C. Ellis. One of the Ellises had participated with Dickson in his early investments in Tonopah.[33]

The mining lawyers for both sides received substantial assistance from the regular attorneys for the mining firms. The Butler's regular attorney was Hugh H. Brown, long-time Tonopah lawyer and one of the most well-connected lawyers in the state.[34] Brown first ventured to Tonopah in 1903, when his San Francisco firm sent him to the desert to help open an office. As a young lawyer, Brown may have encountered W. H. Dickson, top lawyer for the West End, in the courtroom in 1903 when Brown's firm defended the Tonopah Mining Company against the elder lawyer's earlier venture, the Tonopah and Salt Lake Mining Company.[35] Brown did legal work for many of the Tonopah mining firms, especially the Tonopah Mining Company and related companies such as the Tonopah Belmont. Rounding out the Butler legal team was J. H. Evans, Brown's Tonopah partner.[36]

The other West End lawyers collectively represented a wealth of experience from across the West. Harry Hunt Atkinson was a well-connected local attorney. He had moved to Tonopah in 1906, just as he was beginning his career. He ate his meals at the private Nyco Club, where John Chandler, superintendent of the MacNamara and (later, during the period of the trial) the West End, was also a member. Atkinson also knew Chris Zabriskie, one of the principal owners of the West End, "very well." Atkinson even had some family ties to the mines; his father-in-law Clyde Jackson was an early manager of the MacNamara. Elected justice of the peace in 1908, Atkinson served two terms until 1912, and began a two-term stint as Nye County District attorney in 1917. Atkinson's office was in the Nyco Building, where his friend Chandler shared space with Mark R. Averill, who later was elected district judge, and who in 1914 stood ready to rule on the case between the two mining companies.[37]

Horatio Alling was living in California when the trial began, but had lived in Tonopah from 1906 to 1910. He had a reputation as an excellent trial lawyer and continued to work in courts throughout Nevada. "Judge" S. S. Downer made his legal reputation in Boulder, Colorado, serving as county judge, district attorney, and district judge for nearly three decades before moving to Reno in 1904 and joining one of the largest firms there.[38] The West End also used the services of Peck, Bunker, and Cole, of Oakland and San Francisco, probably its regular legal firm (the West End's corporate headquarters were in Oakland). James F. Peck, the senior partner, was well known in the Golden State and earned a specialty reputation with his involvement in water rights controversies in the central valleys. Not only had junior partner Walter D. Cole lived in Tonopah from 1906 to 1910, but also the Nevada Supreme Court appointed him to the commission that compiled the state's laws.[39]

The two companies engaged top experts in geology and mining as witnesses. The Butler's top geological witness, John Wellington Finch, had served as Colorado state mineralogist and as a mining engineer in Cripple Creek.[40] Finch gained fame in central Nevada for managing George Wingfield's Goldfield Consolidated Mining Company, and was credited with being the expert whose advice was responsible for the organization of the Goldfield Consolidated. He also had a reputation for standing as an excellent witness in mining cases, handling long cross-examinations with aplomb, and bearing "the poise born of absolute knowledge of facts."[41]

Finch's career after the trial further enhanced his image. From 1930 to 1934 he served as the dean of the College of Mines at the University of Idaho, then was tapped to head the US Bureau of Mines, under Secretary Harold Ickes of the Department of the Interior from 1934 to 1940 (despite an initial flap by New Deal Democrats over the fact that Finch was a Republican, like Secretary Ickes, and seemed out of place in Franklin Roosevelt's administration).[42]

Another Jim Butler expert, Fred Searls Jr., was also affiliated with Wingfield's Goldfield Consolidated, as a geologist on the payroll for three years and as a consulting geologist afterward. A 1909 graduate of the University of California who had studied under Andrew Lawson, Searls' young geological consulting career was just beginning to take off. On the stand, he mentioned his work for the "Gunn-Thompson people" as well as his other consulting engagements. Almost a decade later, in 1925, Searls would join the firm newly formed by the "Gunn-Thompson people," Newmont Mining Company, and would become famous as a top executive of industry leader Newmont for several decades. That Searls would serve as a good witness in mining law cases was perhaps no surprise, given his family history. Searls' grandfather, Niles Searls; his father, Fred Searls; and two of his brothers, Carroll and Robert M. Searls, all practiced mining law in California. Fred Searls Jr.'s younger brother Robert worked in Lindley's office as a junior attorney, and appeared in the acknowledgments to the 1914 edition of *Lindley on Mines*.[43]

The Butler also retained Andrew C. Lawson, then acting dean of the School of Mines and professor of geology at the University of California, as an expert witness. In 1888 Lawson earned one of the first doctoral degrees that The Johns Hopkins University granted in geology; Joseph LeConte invited him to Berkeley as a professor. Lawson accepted, and taught there until his retirement in 1928. He chaired a committee of geologists put together immediately after the 1906 San Francisco earthquake whose report was a landmark in the understanding of seismic activity.[44] Lawson's studies of ore deposition made him a valuable witness in apex cases. He worked on many of the cases that Lindley tried, including the earlier defense of the MacNamara against the West End.[45]

The mining and geology experts who testified for the West End were no less distinguished. Their primary geological witness was Horace V. Winchell, a nationally known geologist, cofounder and president of the

Geological Society of America, and son of a famous geologist.[46] Winchell coauthored the first scientific analysis of the Mesabi Iron Range, and was an important expert in a variety of other mining legislation.[47] As described in chapter 3, Winchell helped the Anaconda Copper Mining Company set up their geological department in order to prepare for apex litigation, and with David W. Brunton created the innovative Butte System of geological mapping. Winchell started his own consulting business in 1908, having left Anaconda in 1906, and testified for clients all over the world, despite his own well-documented misgivings about the wisdom of the apex law.[48]

The West End also used testimony from Walter H. Wiley, a well-respected mining engineer who in 1883 had been one of the first graduates of the Colorado School of Mines. By the time the trial commenced, Wiley had a thirty-five-year career in mine examination and litigation worldwide.[49] The West End also retained Edmund Juessen as an expert witness. The forty-six-year old American-born Juessen learned mining engineering at Freiberg, and received a doctorate of natural sciences at Zurich in 1890. Like many engineers of his era, he worked at a series of mines in the West early in his career, including a two-year stint as manager of the Pittsburg[h] Silver Peak Gold Mining Company, at Blair, Nevada, near Tonopah. After resigning in 1911, Juessen moved to the Bay Area and worked as a consulting mining engineer.[50]

Both companies rounded out their team of experts with locals who knew the disputed spaces intimately. During the trial individual miners were called to testify briefly, but only one local expert on each side testified at length to local conditions. The Butler team hired a local expert with an excellent reputation: Fred Siebert was a longtime resident mining engineer of Tonopah, for whom one of the major faults in the Tonopah District had been named. Siebert had held many technical positions in various Tonopah mines, including a stint as manager for the Tonopah and Salt Lake property in which Dickson was a major investor.[51]

The West End's local expert, John W. Chandler, had extensive experience with disputes over Tonopah veins. Thirty-eight years old when he took the stand in 1914, Chandler had graduated in 1901 from the Colorado School of Mines, and lived and worked in Tonopah from about 1904 to 1910.[52] For much of that time, Chandler was superintendent of the MacNamara Mine, which adjoined the West End to the north. In 1908 the two companies discovered that they each had an apex claim on a vein that dipped shallowly into the property of the other. (The "north dip-ping vein" in the West End's case against the Jim Butler was the vein that dipped northerly into the MacNamara.) Both sides did extensive work in preparation for a trial, but a late compromise averted actual litigation. The deal, which mining historian Jay Carpenter judged to be more favor-able to the MacNamara, forced the two companies to respect their mutual side line as a vertical boundary. The following year, in 1909, Chandler's MacNamara followed the same north-dipping vein northward out of its claim into the ground of the Tonopah Extension. The MacNamara and the Tonopah Extension prepared to fight in court, but as with the earlier West End controversy, the MacNamara secured a compromise. This time, the MacNamara gave up its apex right in exchange for the Tonopah Exten-sion yielding its right to triple damages (permitted under Nevada law for mining trespass) on the ore the MacNamara had already mined, and both sides agreed to respect the vertical boundary.[53] In the 1908 case against the West End, Chandler worked closely with Lindley and Lawson to prepare the MacNamara's defense, but in 1914, Chandler found himself on the other side. Chandler had held, in his earlier work with Lindley, that there was indeed one vein and that it was an anticline, but in the context of his work for the West End, he had to espouse the two-vein theory. He justified his reversal on the grounds that additional development work proved his earlier statements wrong, but the West End attorneys tried to encourage Chandler to say that he had been coached to see a single vein in the earlier case. Lindley dismissed this attempt in a huff by pointing out, "Certainly he knows as everybody knows that [the single vein] has always been my position and I have not changed it either."[54] Chandler returned to the dis-trict and was hired as superintendent of the West End on October 1, 1914, after preparations for the Jim Butler trial were already well under way.[55]

The trial commenced on Monday, December 7, 1914, and the court began taking testimony the following day. The lawyers and experts addressed themselves only to Judge Mark R. Averill, elected to his seat in the 5th District of Nevada in 1908. Averill was familiar with at least some of these experts and lawyers, and they with him, as Averill had once been among them on the other side of the courtroom. A native Nevadan, born and raised in Virginia City, Averill received at least a little formal educa-tion in mining before serving fifteen years as a public school administrator.

Averill's legal career began in Tonopah in 1903. He served closely with Lindley and Chandler as part of the legal team for the MacNamara as it successfully fought off apex threats from the West End and the Tonopah Extension.[56] The judge had at least a speculative interest in some Tonopah mines. In 1910 Averill, together with two Tonopah bankers, organized the Tonopah 76 Mining Company to coordinate the development of three claims located west beyond the ground in dispute in the case between the Jim Butler and the West End.[57] Despite the cozy connections between Averill and some of the participants on both sides, the Butler and the West End agreed to avoid the additional complexity, uncertainty, and expense of a jury trial, and to have Averill alone pass judgment.[58]

The burden of proof of the asserted apex was on the West End, despite the fact that in this particular suit it was the defendant in the case. Technically, it was the defendant in a lawsuit brought by the Jim Butler to prevent the West End from mining in Butler ground.[59] The West End claimed that there were in fact two separate veins, both of which apexed within the limits of the West End claim, giving the company extralateral rights in both directions. The company based its claim largely on the structural geology of the area. The south-dipping vein, which was the one that dipped into the Jim Butler Company's claim, was slightly older than the other, north-dipping vein, according to the West End's geologists. Furthermore, the West End claimed that the south vein apexed against the footwall, or bottom, of the north vein. This north-dipping vein, the West End contended, continued beyond the juncture with the south vein, and came to a different, independent apex against the overlying Midway andesite or Fraction dacite-breccia cap rock. They also claimed that the Siebert Fault constituted the hanging wall of the north vein.

The West End based their legal case on the strategy of proving that the ore was in the form of two separate veins. The Leadville cases described earlier had set the important precedent in the 1880s that an ore formation that consisted of a folded bed, with synclines (U-shaped formations) and their opposite, anticlines (upside-down U-shaped formations), undulating through the formation, did not have a true apex, and therefore did not have extralateral rights. If the geological formation under the West End were an anticline or a simple undulating vein, then it would have no apex, and therefore no extralateral rights. As a result, the West End's strategy from the beginning was to prove that the two sides were actually two veins. They formalized their argument by providing six specific reasons why the court should interpret the geology as two veins:

1. Difference in the dip of the two veins

2. Difference in the strike of the two veins

3. Difference in the thickness of the two veins—the south vein was thicker than the north vein

4. Difference in the hanging walls of the two veins

5. Difference in the values of the ore, proving that it was not consistent all the way through

6. The upper part of the north vein went above and beyond the apex of the south vein

The Jim Butler's case for denying the West End's apex assertion was to identify the overall vein structure as a single vein with an anticlinal roll.[60] If it was an anticline, the Butler attorneys argued, then no apex was present, and therefore the West End violated the rights of the Jim Butler when it followed the south vein into Butler ground. The Butler team replied to each of the West End's six justifications in turn. To the assertion that the dips were different, the Butler replied that the structure in question was a roll, which meant that dip was irrelevant. To the question of strike, the Butler pointed out that if the axis of the anticline deviated at all from horizontal, then the strikes of the limbs of the anticline must be different. The Jim Butler replied to the West End's assertion that the south vein was thicker than the north vein by noting that there was not, on average, any difference, and that both limbs varied considerably in thickness. To the West End's fourth reason, that the hanging walls were different, the Butler team argued that the hanging wall was the same, except where it was cut off by the action of the Siebert Fault. (The action of the Siebert Fault in severing the roll of the vein can be seen in figure 5.3.) The Butler tackled the issue of differing values of ore head-on by contending that valuable ore continued over the top of the roll of the anticline, and further that excavations of ore (called stopes) went over the top of the roll under a continuous hanging wall—a clear sign of an anticline. To the West End's final charge, that the north vein continued upward beyond the apex of the south vein, the Butler replied that the quartz above the anticlinal roll was a mass of stringers that took the form of a "halo" above the roll, but that

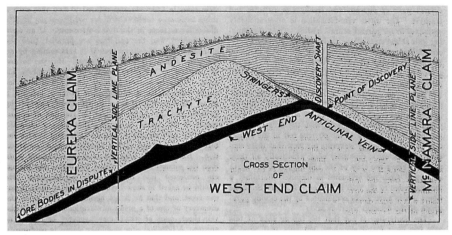

Figure 5.4. Cross-section of the disputed area, showing stringers, vein, and surrounding rock types. From Robert M. Searls, "Apexes and Anticlines," *Mining and Scientific Press* 117 (1918): 43.

did not constitute an extension of the vein. The projecting stringers are clearly labeled in figure 5.4.

In the formal arguments outlined above, both sides were careful to use language that supported their vision of the underground. The West End always spoke of the "north vein" or "northerly dipping vein" and its southern counterpart, even when the two came together underground. The Jim Butler team was similarly careful to always describe the "anticline" or the "roll," instead of the apex, and usually termed the two sides of the anticline as "limbs" or "branches" instead of the West End's favored "veins." The Butler had a little more leeway here, since it believed that there was just a single vein in the form of a roll; the West End was careful to always specify the north-dipping vein or south-dipping vein whenever discussing geological conditions.[61]

In order to convincingly argue their points, both the West End and the Jim Butler utilized the visual culture of mining to the best of their ability, creating a wide variety of visual representations of the underground spaces in dispute. The most expensive and elaborate of these were the models made by each side. The models looked quite different and were intended to play different roles in the legal strategies of the two companies.

The West End legal team, led by Dickson, opened the trial by introducing their visual representations.[62] First to take the stand was George W.

Pierce, a model maker from Los Angeles who had constructed the West End's model at a scale of forty feet to one inch on the basis of maps and drawings provided by the West End engineers. Pierce worked constantly from July to December 1914 on the model.[63] The model itself purported to show the true state of the underground, minus some inaccessible workings, as of November 1, 1914.[64] Richardson, the mining engineer from the West End, periodically checked Pierce's work for accuracy, and he was the next to take the stand. Dickson used Richardson to explain the first seven defense exhibits. Exhibit A was the three-dimensional model, made by Pierce, which figured so significantly in the proceedings to come (see figure 5.1).[65] Next were five maps of the West End and adjacent workings, platted horizontally. Each map represented a horizontal level of the mine, like a layer in a cake, seen from an overhead perspective. The last map consisted of four representations of vertical sections through the mine. These section lines were also marked on the horizontal maps. By the time Dickson turned Richardson over to Lindley for cross-examination, the West End had given the judge ten visual representations of the West End ground—five horizontal slices, four vertical slices, and one complete three-dimensional model.[66]

The West End's three-dimensional model of the underground played a central role in their overall legal strategy. The large skeleton model served as a three-dimensional key for all of the visual representations used by the West End, and it also gave expression to rhetorical discipline comparable with the lawyers' careful efforts to always refer to separate veins.

The model, which the West End team referred to throughout the trial as though it was simply factual evidence, actually embodied the West End's arguments about the geology of the disputed vein. The most powerful arguments were made by the choice of paint colors. Though the rock in the vein (or veins) was essentially identical, the West End painted the south vein a bright red on the model, and painted the north vein a vivid yellow. This is particularly noticeable when the two veins come together, as shown in figure 5.5. The arbitrary color choice is made more clear by the fact that they also chose to paint the fraction vein, a third vein or branch not of direct consequence in the suit, the same red as the south vein. Colors were also used to emphasize arguments about geological distinctions. For example, on the model the trachyte rock was colored purple, and the andesite, which formed part of the cap rock, was a light green.[67] However, the two rocks that the widely different colors represented actually looked virtually identical to the naked eye—the distinction was justified only on the basis of slight differences visible in carefully prepared slides under the microscope. Geological experts who studied the district disagreed on even the presence of all of the rocks, much less their proper names.[68] Yet the West End model made a stark distinction between two very similar rocks with uncertain origins, because such a distinction supported their theory of the formation of the ore, which in turn supported their two-vein distinction for legal purposes. The rock from the veins, so clearly distinguished as red and yellow on the model, was identical-looking quartz. Even the West End's experts admitted the impossibility of telling apart samples from the two veins if they had been removed from the mine; the only difference was structural, in the direction they were oriented in the ground.[69] The West End's two-dimensional maps and geological sections also conformed to the same scale, numbering, and color scheme, providing a unified and consistent chromatic argument for the truth of the West End's geological assertions.[70]

During the testimony of their expert witnesses, the West End legal team validated the model first, then had the expert describe the underground, while referring to the model or sectional maps whenever possible. Dickson guided Juessen's early testimony in this way:

Q. You are familiar of course with the coloring on that model?

A. I am.

Q. Do you from your examination endorse the coloring there which is intended to represent the south vein throughout, as correct?

A. The coloring—The red color represents the south dipping vein, or the West End vein.

Q. Does it correctly represent it according to your examination of it on the ground?

A. It does.

Q. And state whether or not on your examination you endorse the coloring of the north dipping vein as shown upon this model, and upon the other exhibit?

A. I do.

Q. And do you also endorse the coloring on this model, upon the cross section, Exhibit G, which are intended to show the enclosing rock,

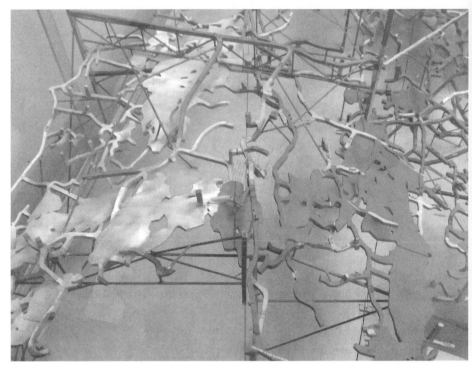

Figure 5.5. This view from overhead of the West End's model shows one place where the "north-dipping vein" and the "south-dipping vein," painted yellow and red, respectively, came together in a continuous stope over the top of the anticline. But the company chose contrasting colors to reinforce its interpretation of the structure as two separate veins that happen to meet at the top. Model in the W. M. Keck Museum at the Mackay School of Mines, University of Nevada–Reno. Photo by Eric Nystrom, 2012.

or foot-wall and hanging-wall boundaries of the two veins wherever they have been ascertained?

A. I do.

Q. Is there anything, any feature, any point in the color scheme as adopted in the preparation of that model which does not meet with your approval?

A. There is none.[71]

Dickson was also careful to integrate the model into the testimony on the underground. For example, with Juessen on the stand, after the expert testified to his satisfaction with the model, Dickson had the witness stand up and move over to the model itself. The lawyer then asked a series of questions about each raise into the geologically disputed ground, using the model as a reference. At one point Dickson asked a series of questions about raise 5-A—the amount of quartz (which contained the vein), faults, and so on:

Q. How far up the raise does the quartz continue?

A. The quartz continues 71 feet, up into the raise and terminates against a small fault.

Q. How far is that, if at all, below where the raise as shown upon the model must have encountered trachyte and cap?

A. Trachyte and cap were encountered 271 [feet] up in the raise above the hanging wall of the south dipping vein at a point indicated here by the purple in the raise immediately below the green, indicating the cap.[72]

The company carefully used the West End model as a visual key to the testimony of the other expert witnesses it called. When Dickson brought John W. Chandler to the stand for the West End, the lawyer and the witness used the model to organize Chandler's testimony. After being introduced, Chandler produced a rock, taken from the vein, and described where it was found in the mine, in reference to the model. Thus the West End model was now associated with a factual geological specimen, even though the sweeping color scheme was suggestive of a far greater uniformity than was actually found underground. Next, Dickson had Chandler testify about the conditions he found in the area where the apex should have been. In the order they were found on the model, Chandler pointed to a raise, described what he had found in the mine at the point represented on the model, then moved to the next and did the same.[73] After finishing with the raises, Chandler used the model to describe how the two veins, in the West End's formulation, butted up to each other. (The model's representation of a portion of this area is shown in figure 5.5.) At one point Chandler used one of the West End's two-dimensional sections, Exhibit G, to illuminate a portion of the ground that was difficult to see on the model, but the section used the same color scheme as the model and the expert returned to the three-dimensional artifact as soon as it was possible to do so.[74]

The Jim Butler team, in contrast to the West End's strategy of portraying the underground with a single skeleton model plus two-dimensional maps and sections, created and deployed three different models, plus maps and sections, to make its case. Unlike the West End, whose skeleton model served as the key to all the other two- and three-dimensional visual representations, the Jim Butler team's Exhibit 1 was a large horizontal (bird's-eye) view composite map of the underground workings of both the Jim Butler and the West End, as well as parts of the MacNamara and Tonopah Mining Company. The map was made to a scale of twenty feet to an inch from surveys conducted by E. C. Uren, who oversaw the construction of the visual representations for the Butler team, as well as two other surveyors. (Uren had drawn most of the illustrations for the third edition of *Lindley on Mines*.)[75] Except for the stopes, which were shown in a brownish-yellow, the workings on the map were colored to distinguish one level from another, though not with any specific key, unlike the models. The boundaries of the two mines were also outlined, the Butler in red and the West End in green. Uren numbered his surveyors' stations, and used West End numbers whenever his crew followed the other surveyors, though this practice created some numerical discrepancies when the Butler engineers visited a place first, since the West End surveyors did not use the Butler numbers. He also distinguished between raises and winzes by including an appropriate up or down arrow. The mapmaker put additional "arbitrary" numbers in strategic spots on the map, such as at the ends of drifts or "at points where I thought testimony would be introduced." The map that formed Exhibit 1 also had lines representing vertical sections through the workings, which served as other exhibits.[76]

The Butler's own skeleton model served as Exhibit 2. It was made from data from the surveyor's notes that were the basis of the large map of Exhibit 1, and was the same scale (forty feet to an inch) as the West End model. Mr. Douglass, of the Tonopah Belmont Mining Co., had constructed the skeleton, with the supervision and assistance of Uren. The vein had been painted red, with the Mizpah trachyte green, the Fraction vein brownish orange, the Fraction dacite breccia yellow, Midway andesite purple, the glassy trachyte of the Tonopah Mining Company light blue, and the West End rhyolite brown; faults were depicted with blue stripes. The geologists Finch and Searls had directed the coloring of the various parts of the model.[77]

Exhibit 3 for the Jim Butler team was a glass section model. Uren constructed the glass model himself, working from geological sections made by Finch and Searls; he "checked and re-checked" it for accuracy.[78] The thirteen plates corresponded to exact north-south sectional lines, one hundred feet apart, through the most important part of the anticline. The section lines were also clearly marked on the large map, Exhibit 1. The glass sections showed the veins, as well as faults, and the lines of the drifts, levels, raises and winzes. If the projected section cut a mine excavation exactly, the level appeared outlined in black, but the section also showed nearby workings outlined in a dotted line. The geology was colored or filled in: the Fraction dacite breccia was outlined in black and tinted light green; the Midway andesite received a black cross-hatching; the Mizpah trachyte was depicted with green dashes; and the West End rhyolite was indicated by a lines-and-dots pattern. Uren's chromatic palette may have been hampered by the need for each glass section to be at least semitransparent, so viewers could also see the glass sections behind it.[79]

The final model used by the Butler team, introduced as Exhibit 4, was a block model, made by Uren, again under the direction of Fitch and Searls, to a scale of one hundred feet to an inch. The complete model showed the present surface of the ground. The top piece could be lifted off, to allow the model to show the Trachyte Surface, which was the surface of the earth immediately before the lava flows turned it into the present surface. Removing the second piece revealed a third surface (dubbed the Original Surface in Averill's published opinion, for lack of a better term), which depicted the ground as it appeared immediately after the formation of the vein and before erosion converted it into the Trachyte Surface. The Original Surface, according to the Butler team, was the oldest, followed over a relatively long period by the Trachyte Surface.[80]

After introducing the large map and the three models, briefly describing them, and testifying to their concordance with his surveyors' notes, Uren's time on the witness stand should have been done. The strategy that both sides agreed to follow involved allowing the geologists to introduce the more specific maps as necessary during their testimony, and saving the interpretation of the visual representations for the expert witnesses. For example, Lindley asked only two short questions of George Pierce, the West End model maker.[81] However, with Uren now on the stand, Dickson aggressively attempted to sow doubt about the Butler models.[82] The lawyer tried to have the West End model moved immediately next to the Butler skeleton model, but Lindley blocked the maneuver. Dickson's cross-examination strategy was to push Uren to admit that the West End and Butler skeleton models were "practically identical," nearly the same except for color. Then Dickson asked about some minute differences—variations in a particular stope, and the absence of some workings (in the Butler ground) from the Butler model, which Uren admitted. The lawyer set up some tricky questions, asking whether the Butler model portrayed certain stopes in Tonopah Mining Company ground, for example. Uren replied that it did, and Dickson then pointed out that the West End model covered that ground too, plus more beyond where the Butler model stopped. Dickson also asked Uren, the precise model maker, a series of broad, sweeping questions—whether the model represented all the workings in the claims of both companies, for example—which Uren had to deny, then attempt to clarify so as not to look foolish. These questions were intended to establish that the Butler skeleton model was less comprehensive and less factually reliable than the West End model. Then, Dickson turned to sow doubt about the glass model, the Jim Butler's Exhibit 3. He began by invoking the realism of the skeleton model:

Q. So if the surface overlying the vein or veins has been removed, and the earth from that on down was transparent as glass, anyone walking over it and looking down could see the workings and the stopes on the ore, just as they are in the mine?

A. Yes.

Q. That being so, what additional light is thrown upon the condition, as you understand it, by the . . . glass model?[83]

Didn't the skeleton model have all of this information and more?, reasoned Dickson. Uren pointed out that the faults were easier to see on the glass model, but the real value of the glass model was that it was easier to visualize the structure of the geology. But wouldn't anyone looking at the skeleton model be able to "have a perfect picture without projections?" asked Dickson. If the skeleton model was complete, wouldn't it have the information? Uren discussed the veins that had not been excavated, which were visible on the glass model, and Lindley tried to intervene, but Dickson doggedly returned to the comparison that the model maker was attempting to avoid.[84]

Q. Is there anything represented on the glass model or any of the several sheets of the glass model, that could not be observed by one if he was looking through or along that identical plane of section under the ground; that he could not see if he had this wooden model before him?

. . .

A. Not as to actual openings, no.

Q. Nor the ore occurrences, so far as they have been developed?

A. So far as they have been developed.

Q. That is all.[85]

From the very start, then, Dickson had shrewdly created doubt about the Jim Butler's visual representations by comparing them unfavorably against one another. He also added to the West End model's authority by making it clear that it covered more terrain with a greater attention to detail, and was suitable to examine where any model was needed. Dickson tried to ensure that if the two models conflicted the West End's skeleton model would be perceived as more factual.

The Jim Butler team's visual representations portrayed a less cohesive strategy than that employed by the West End. The West End had a single, key model, with maps and sections drawn to the same scale and employing the same color palette. The Jim Butler, by contrast, used three models, each with different scales, and a large map scaled differently from the most important models. The map and the glass model were both twenty feet to an inch; the Butler skeleton model was forty feet to an inch (as was the West End's model); and the block model was one hundred feet to an inch. The Jim Butler's visual representations also did not correspond with each other chromatically in a consistent fashion. On the large key map, levels were colored almost at random, with the intent only of distinguishing the levels from one another, yet on the skeleton and glass models the same colors were used to represent different geological or mining facts. Additionally, the models did not use the same colors for the same things. To be sure, the models were intended to highlight different aspects of the phenomena under discussion, but the lack of chromatic unity certainly denied the Jim Butler side the subtle rhetorical authority of the West End's exhibits. The differences between the visual representations provided Dickson an opportunity to sow doubt about the reliability of all of them, while implicitly increasing the perceived reliability and correctness of his own side's models and maps.

To the Jim Butler team, concerns about the need for consistent scale or colors were relatively immaterial. During Dickson's interrogation of Uren, when the West End lawyer was trying to push Uren into admitting that the glass model was a redundant display of information already on the skeleton model, Lindley's interruption of the exchange revealed his thoughts on the appropriate uses of visual representations in the courtroom: "[A]s to the comparative value of these two models as evidence, I do not think that it is proper cross examination. Of course, I do not care—they show for themselves, and we all know that these models are made to illustrate testimony as much as anything else."[86]

Lindley thought of the visual representations as illustrative of testimony and arguments, not so much as factual evidence. For Lindley, maps and models were intended to help the viewer understand the true nature of the underground. Showing veins and structures that had not been excavated, or that had once possibly existed but had eroded or been covered (and thus, in both cases, now had to be imagined) increased the value of the work as a representation. The limits of the plaintiff's imagination were bound by the evidentiary rules of science of geology, which permitted (or even required) at least a little cautious inference and speculation. Lindley's visual representations were not intended to be read in a more narrow way as strictly true or false evidence.

The testimony was concluded on December 22, 1914. On March 8–11, 1915, the lawyers orally argued the case before the judge, and on April 30, 1915, Averill issued his verdict. The maps and models made by both sides appeared to have made a significant impact on Averill's thinking, since he referred to them many times in the decision. Averill's decision in the case, and his reasoning behind it, came as a bit of a shock to those who had been following the proceedings. He began his opinion with a recognition of the importance of getting the geological facts straight in his opinion, since higher courts would almost certainly revisit only points of law, not points of fact.[87] He commenced his discussion of the case by declaring that there was indeed a single vein, but, Averill cautioned, "[T]his conclusion is not at all what might be called an absolute one, but is qualified, as will appear further in this discussion." Averill said that the vein existed in an anticlinal form, but "strictly speaking it is not true geometrically or geologically." The judge then quoted *Lindley on Mines* that, geometrically speaking, an anticline has one or two synclines, but that was not the situation in this case.[88] Averill offered an additional caveat about the difficulty of determining geological facts, and recognized that "whatever result may be reached in this case, it must be based upon an incomplete presentation of evidence. Should it develop in the later mining history of the Tonopah district that the conclusions of fact in this case are wrong, the error should be excused for this reason."[89]

Averill thus prepared himself for a conservative ruling. He declared that not enough work was done in order to show the complete anticlinal roll. Instead, its junction was proven to exist in only a few places. Averill described in great detail the edges of the veins, as far as they were known, and also described the term *anticlinal axis* he had invented, which was an imaginary line connecting the highest points along the top of the anticline where the two veins were proved to have merged. This anticlinal axis then could stand in for a line of an apex.[90] The judge said the apex lines were "obvious," which he contrasted with a newly invented antonym "subvious" that had been used during trial arguments. Perhaps sarcastically, Averill pointed out, " 'Obvious' is a good word."[91] "Obvious" reasoning, not coincidentally, was well supported by the visual arguments made by the West End model. Throughout the rest of the opinion, Averill referred to the model extensively, along with one other diagram he created himself, to explain his argument. Averill handmade his two-dimensional diagram by measuring the West End model with a ruler, and it was intended to represent the highest points of the veins, with the direction in which they dipped(figure 5.6).[92]

There may also have been confusion at this point about the burden of proof. The burden of proving that a clear apex exists belongs to the company attempting to assert extralateral rights, which in this case was the West End. However, because of the particular legal maneuvers that preceded the case's trial, the Jim Butler was listed as plaintiff and the West End as defendant, even though the burden of proof of an apex was still on the West End. Averill makes note of this fact, that one possible reason the ground was left unexplored was because the burden of proof of an apex was on the West End, so the West End did not want to explore and find an anticline, and the Butler was content that the West End would not have been able to prove the existence of a double-apex.[93] Averill noted, however,

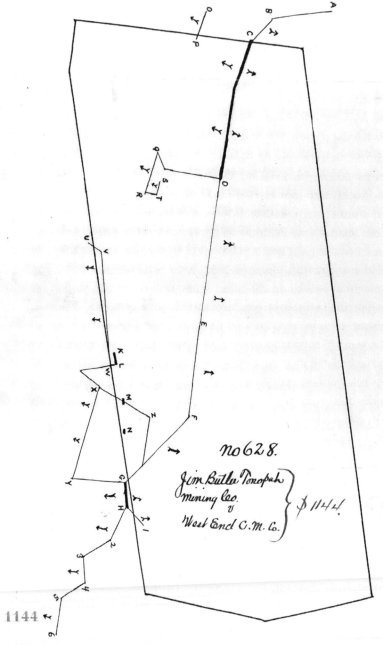

Figure 5.6. Judge Mark Averill made this diagram on his own by measuring parts of the West End's skeleton model with a ruler. It depicts the outline of the West End claim, as well as an "anticlinal axis" that the judge believed could stand in for a traditional apex. *Jim Butler Tonopah Mining Co.* v. *West End Consolidated Case File*, US Supreme Court Appellate Case Files, 25458, Number 249, Box 5000, Record Group 267, National Archives, Washington, DC, p. 1144.

that the Butler contention was that the ground looked like the Butler's block model, which showed the roll in a continuous fashion. Therefore, Averill clearly believed, in phrasing his argument about the Butler not doing enough work to prove the anticline existed, that they had not gone to enough trouble to uncover geological facts to back up their assertions. Later in the opinion he described the Butler block model as "open to objection" (despite the well-publicized fact that neither side brought any objections at all during the trial) and noted, sarcastically, "[A]ll this is ingenuous as well as artistic."[94] He compared the Butler block model implicitly to the West End skeleton model, which showed the truthful gaps in the underground, and found the block model deficient as evidence: "The vein does not exist as one unbroken sheet as depicted on the [Butler block] model, not, at any rate, as shown by present development. If, therefore, the Butler contention rests upon this model, it rests upon a foundation consisting largely of assumption, which, however, might have been turned into fact in part by proper development of the 'Terra incognita.'"[95]

Averill then tackled the halo issue. He opined that the halo was more extensive, and that, as a continuation of the vein, must have reached the Trachyte Surface (which was the surface after the deposition of the vein but before it was all covered over by dacite from the eruption of Mt. Brougher). The fact that the halo was part of the vein and reached a prehistoric surface gave the West End a terminal edge on the vein, traditionally a key part of determining an apex.[96] Averill admitted that the quartz of the halo, where it would have met the surface, was not oxidized, which was a Butler argument to counter the idea that the vein had ever reached any prehistoric surface. Averill waved off the concern by noting, "[H]alf a dozen explanations can be offered against any such conclusion, one of which is that unoxidized veins are often found that undoubtedly reached a surface when forming, perhaps some in the Tonopah district." Furthermore, Averill pointed out, this reasoning only needed to apply to 240 feet of the juncture of the two veins, since this was the only part of the united vein that had been proven to exist.[97]

Averill then briefly discussed the issue of the roll of the vein. The judge argued that the geological section offered by the Butler team as proof of the vein's roll was misleading—not because it was drawn poorly, but because the particular part of the vein shown by the section was not representative of the vein as a whole. Here Averill is again implicitly comparing a localized vision, provided by the Jim Butler's geological section, with the broader and more obviously "factual" vision provided to the judge by the West End model.[98]

The judge concluded his discussion of the geological conditions by addressing the strike of the vein. The Butler had argued that what the West End claimed was the dip of the vein was actually its strike, and vice versa. The lawyers saw this as important, because extralateral rights followed the dip and not the strike. If the Butler's argument was true, it would probably mean that the apex of the vein (if one existed) was not even in the West End claim at all, and that the property line with the Butler would have to be respected as a vertical plane. At minimum, if the dip was actually the strike and vice versa, the West End would not have proven the existence of an apex, and no extralateral rights would exist. Averill noted that the "mathematical" strike of the vein depended largely on which points were used to calculate it, and that the strike could vary wildly. However, argued Averill, "as a matter of common sense rather than technical mathematics the strike of the vein as a whole is easterly and westerly, and the strike of its two slopes . . . is as claimed by the West End."[99] The West End model would have facilitated Averill's "common sense" decision here, mostly because it left out the territory necessary to prove the Butler's contention. Indeed, a quick look at the model does make it "obvious" that the strike of the two veins or branches conforms to the West End's argument.

After discussing the geological arguments, Averill turned to the legal implications of the case. Three legal points remained to be settled. The most important question was whether a vein could have extralateral rights in two directions. Averill said that it depended on the specific facts of the case, and in this case it was possible to find a terminal summit at the crest of the anticline, which made "that line of highest points a true apex for the whole vein, from which apex the vein can be followed downward both northerly and southerly."[100]

Averill also made short work of the second Butler argument about the law. Lindley had argued that since the extralateral rights the West End was trying to claim were on a vein other than the one that the West End had discovered originally (and that was, therefore, the basis for their right to mine the ground), they could not have extralateral rights on the other vein. Averill judged that Lindley's argument was "fully and conclusively answered" by the West End, who argued that clear court precedents and even language in the 1872 Mining Law gave mining companies extralateral rights on any and all veins apexing in their claim, not just the so-called discovery vein.[101]

Averill then tackled the third legal point, which was a dispute over the end lines of the West End claim. In an ideal rectangular mining claim, the end lines were the short sides of the rectangle. The long sides were called side lines. Such an ideal claim would be laid out over the apex of a vein, so that the vein bisected the rectangular claim lengthwise. Accordingly, the ideal vein would cross both end lines of the ideal claim. These end lines were projected vertically downward, and limited the rights of the mining claim. Miners could follow the vein extralaterally beyond the side lines, but not beyond the end lines. The mining law of 1872 held an additional restriction on the end lines—it required them to be parallel. (This made sense, because nonparallel end lines would either converge, limiting the ability of the claimant to go extralaterally beyond the side line, or diverge, giving the claim owner more ground extralaterally than was owned by the original claim.) The principle of parallel end lines limiting the boundaries of extralateral rights was thus fairly simple, in idealized situations. Lindley argued that the western end line of the West End claim was not parallel to the other end line—in fact, the western line was jagged, a bit like a lightning bolt. Thus, according to Lindley, the claim was not laid out according to the requirements of the law; if that was true, then the West End could have no extralateral right.

However idealized the original law may have been, geological situations were rarely ideal, since veins often meandered out a side line instead of the end line. That is what happened with the West End—the discovery vein came in through the eastern end line, turned, and crossed out through the northern side line instead of the other end line. In an earlier case with a meandering vein of this sort, the US Supreme Court had ruled that the appropriate solution was to draw an imaginary vertical plane, parallel with whatever end line the vein passed through, at the point where the meandering vein crossed the side line. The claim owner then possessed extralateral rights between the end line and the imaginary plane.[102] On the basis of this reasoning, Averill decided that the jagged western end line of the West End had no bearing on its extralateral rights. Averill did note that, technically speaking, it was necessary to have a valid claim before any extralateral rights could exist, and one of the four criteria for a claim's validity was that it must have parallel end lines.[103] However, Averill pointed out that later courts had interpreted the requirement rather loosely, and suggested that the West End's lines were close enough to preserve the validity of the claim.[104]

Averill realized that his legal conclusion—that the vein was an anticline, but that an apex still existed and that extralateral rights could flow in opposite directions from the same claim—was radical and perhaps even counterintuitive. Averill attempted to forestall criticism by quoting from a noted mining judge of the previous century, Chief Justice of Nevada William H. Beatty. "If the conclusion reached herein seems revolutionary," wrote Averill, it would be wise to remember Beatty's argument for judicial flexibility: "We are willing to admit that cases may arise to which it will be difficult to apply the law; but this only proves that such cases escaped the foresight of Congress, or, that although they foresaw the possibility of such cases occurring, they considered the possibility so remote as not to afford a reason for departing from the simplicity of the plan they chose to adopt."[105]

In what may have been a subtle insult directed at the Jim Butler's counsel, Averill lifted the Beatty quote—verbatim and without attribution—from the same section of *Lindley on Mines* that Averill used earlier in his opinion to justify ruling that the vein in this case was not a geometrical anticline.[106] With his factual and legal conclusions completely explained, Averill ruled that the West End owned the apex, and was therefore entitled to follow the vein out both of the claim's sidelines, including into Jim Butler territory.[107]

It is clear from Averill's opinion that the judge did not consider the Butler models and the West End skeletal model to be equally factual. Dickson's early work with Uren, the model maker, on the witness stand had sown doubt about the utility of the Butler skeleton model. The West End model was at least as good, more comprehensive, and more thorough, and could therefore substitute for the Butler skeleton model. Averill pointed out that the Butler skeleton model showed more of a unified apex than did the West End model, but instead of taking this point as a challenge to the West End's story, Averill interpreted it as falsely depicting underground spaces that were not known with certainty.[108] As a result, Averill believed the West End model could be used in place of the Butler skeleton model, with the only consequence being greater truthfulness. The unforeseen consequence, however, was that the West End model's chromatic arguments would no longer be disputed by a countervailing representation.

Next, Averill compared the West End's model to the Butler block model, and concluded that the West End's model provided the superior view of the underground. Indeed, the block model seemed to signal that the Butler team was making rash assumptions about the underground. In the copy of the opinion included with the transcript that was sent to the US Supreme Court, Averill had written a sentence calling particular attention to "a criticism of plaintiff's block model" that he would make later, but that line was struck out and did not appear in the published opinion.[109] The block model made assumptions about the vein that may have been reasonable to a geologist or mining engineer, but that Averill clearly considered to be unproven assertions. It was easy for the judge to superimpose his own knowledge—that the underground had not been completely excavated—on to the Butler block model. That the block model could show a vein to exist where humans had never gone showed that the model was misleading or taking liberties with the truth, at least when the truth was construed in the more rigorous sense used by the legal profession. What passed as proof for a science like geology did not always meet the standard in a courtroom, at least not for Mark R. Averill.

The West End model, especially when contrasted with the Butler block model's assumptions, seemed more believable to the judge. The model gained authority from its carefully engineered origin, to scale, corresponding with other maps and sections, and inscribed with numbers. Its very skeletal messiness seemed to reflect the truth, not to take liberties with it. In short, because of its design it was easier for a person like Judge Averill accustomed to thinking about truth and falsehood in a rigorous, legal way to accept the West End model as fact, rather than as argument. Once the West End skeleton model underwent this transmutation into being a source of fact, the argumentative portions of the model—its color choices, its presentation of stark boundaries where none existed, its orientation and point of view, its inclusions and exclusions—were significantly harder to distinguish from the factual content of the model. It is clear from Averill's written opinion that he relied on the West End model to show the truth, even when some of the truths it portrayed were fictions, the fruits of arguments carefully crafted in a visual language.

The Jim Butler team was clearly unhappy about the "novel and unwarranted" result of the district court trial, though they recognized that some of their arguments were stronger than others. "I have no great faith in the question of the broken end-line, but my opinion is fixed and crystallized that no appellate mining court would ever approve Judge Averill's grant of extra-lateral rights in two opposite direction[s] on one and the same vein," wrote Hugh Brown.[110] The Butler lawyers quickly appealed the case to the Nevada Supreme Court.[111] The court heard the case on November 26, 1915, and rendered a judgment on July 3, 1916.[112]

The Butler team argued that there were two factual problems with the earlier case, and three legal errors as well. They quibbled directly with Averill's finding that the two branches of the vein were separated in the West End claim over long distances, thus forming terminal edges necessary for apexes. The West End's model, however, would have made Averill's view easy to see. The Butler team also took issue with Averill's finding that the quartz in raises driven to explore the halo area was part of a vein.[113] One of the legal challenges argued against extralateral rights on non-discovery veins with alternate dips, and a second challenge objected to the West End's nonparallel claim end lines. The third addressed the anticline as apex directly. Averill had noted that the two limbs of the vein were joined at the roll, and at those points, argued the Butler, he should have denied the West End any extralateral rights. "No extralateral right whatsoever attaches to an anticlinal occurrence in a vein, since the latter does not constitute an apex within the meaning of the federal mining statute," the lawyers argued.[114]

The West End replied by first invoking, in soaring language, the entitlement of "he who brings into circulation the hidden treasures of the earth" to the fruits of such labor, since the resulting wealth "promotes the prosperity of all." They continued in a more specific frame, using extensive citations of prior cases, as well as citations to *Lindley on Mines* and other mining law treatises, to refute the points of law raised by the Butler team. The West End first pointed out that "questions of doubtful construction" should not be automatically decided against the person attempting to exercise extralateral rights, and that quibbles over end lines were certainly that.[115]

The West End legal team devoted most of its energy, however, to arguing about the legality of the vein. First, they attempted to sow doubt about the importance of having a terminal edge to prove an apex. The term *terminal edge* is the fruit of court decisions and does not actually appear in the law itself, noted the West End lawyers. They conceded the term was frequently invoked in apex cases, but that was simply because most veins actually outcropped on a hillside, making them appear to have a terminal edge. But, they argued, "[n]o court has been called upon to determine or define what, within the meaning of the [mining] act, is to be regarded as the top or apex of a vein or lode, having the form of a single anticlinal fold." The lawyers then quickly noted both that the deposit in question was indeed considered a vein, in the context of the law, and that the term *top* in the mining law should be considered "in its popular sense." As a final encouragement to the judge, the lawyers argued, backed by their largest set of citations to other cases, that a "fundamental" rule of law was that any language used by a court decision should be "read, understood, and applied in light of the facts of the particular case with which the court is dealing."[116] In other words, the West End implicitly prodded the court to reject the idea, established in the Leadville cases, that an anticline could not have an apex, because that court precedent should be understood only in the context of Leadville's geology.

Judge Frank H. Norcross weighed the arguments and authored the opinion for the Nevada Supreme Court. Norcross was a "very able Judge, familiar with mining because of his long residence in Nevada," according to Colby's later recollections.[117] The court began by summarizing the facts of the case as it understood Averill to have settled them, including Averill's opinion that the evidence was insufficient, "meager and unsatisfactory," to use the words of the Nevada Supreme Court, that the two branches of the vein actually formed a single "juncture or union."[118]

Norcross then boiled down the case to two key questions—the legality of the West End's western end line and possible repercussions, and the question of whether an anticline can be an apex with extralateral rights. Norcross quickly determined the question of the West End's broken end line: The end line simply looked jagged because the West End claim in its original shape as a parallelogram was staked over an older claim. When the patent was issued, the West End was given all the land covered by its claim, except the corner where the older claim held priority, thus making the end line appear jagged, but this was not, argued Norcross, a reason to deny the West End's ability to exercise extralateral rights.[119]

Norcross then tackled the larger questions about the structure of the vein and the implications of its form on the apex rights. First, was it possible to have extralateral rights in two directions? The Butler team had tried twice to find implicit justification that such rights could flow in one direction only, but these arguments were weak in comparison to the mining law's explicit declaration that a claim owner was entitled to "all veins, lodes, and ledges" with an apex in the claim.[120] Based on this reasoning, Norcross set the parameters for the final solution of the case: "If a vein in the form of a single anticlinal fold may be said to have an apex, we think there is nothing in the statute which militates against extralateral rights upon such vein in opposite directions, the same as though it were two veins with separate apices, instead of one vein."[121]

Norcross then turned to the final, most serious issue: Did the vein have an apex? While no case law provided precedent on this particular situation, many cases had used the term *terminal edge* synonymously with *apex*, thus solidifying the latter concept. In presenting these examples, the Butler lawyers argued that since there was no terminal edge, no anticline could be an apex. Norcross disagreed. He pointed out that the term *terminal edge* was not directly in the law, and the term had not been "of universal use in defining an apex."[122] Instead, it was a convenient term to help translate the terms of the statute (*top* or *apex*), which were not mining terms, in a way that was useful for deciding if an exposed vein was an apex. Norcross pointed out that the intent of Congress, in writing and passing the mining law, was quite clear—mining on the public domain was to be encouraged in every way, which meant that the courts ought to "construe the statutes liberally in the interest of the miner," in questions of extralateral rights.[123] The law, in fact, distinguished only between deposits that were "in place" (i.e., veins or lodes) and those that were not (i.e., placers). Norcross quoted a long paragraph from the "great work on Mines" by "the distinguished counsel for [the] appellant" to the effect that, if a deposit was in place it must be a lode, and if it is a lode, it must have an apex, at least according to the mining law, for there were no exceptions made by the congressional language.[124] Both sides admitted that the feature was a vein, continued the judge, the federal law says all veins are open to location, and every vein must have an apex. Logically, "it follows as a necessary conclusion that the vein in controversy has an apex. If it has an apex, there is no other possible place for it to be than at the crest of the anticlinal fold."[125]

So if it was a vein, in place, it had to have an apex, and if it had an apex, it had to be at the anticline. But what of the terminal edge precedents? Norcross pointed out that standard legal practice involved ensuring that terms were carefully applied to the particular facts, in this case geological, of the case at hand. As a result, the term *terminal edge* appeared as a requirement in cases where they were talking about veins that actually did have a terminal edge. Examining the Butler's precedents one by one, the judge pointed out how each was different, geologically, from the vein at issue in the *Jim Butler* v. *West End* case.

Norcross's discussion of the term *apex* hinged, as the Butler team hoped it would, on the term's use in the Leadville cases.[126] However, Norcross drew opposite conclusions from the cases, prompted in part by the careful explanation of the context of the cases in *Lindley on Mines*. Where the Butler team hoped to draw on the precedent of undulating deposits having no apex and therefore no extralateral rights, Norcross instead chose two quotes from *Lindley on Mines* to emphasize the localized context of the Leadville deposits. There, the ore was a very different sort of deposit, nothing like the vein or veins at issue in Tonopah. The fact that a Leadville anticline had no apex did not mean anything in the context of a Tonopah anticline.[127]

Having addressed the legal issues of the case to his satisfaction, Norcross concluded his opinion. The West End vein had been correctly claimed. If every vein, under the law, must have an apex, then the crest of the anticline must be the apex of this vein. Since there's no question that the apex of the vein, wherever it might be, does not fall in Jim Butler ground, it is clear that the decision made by the lower court was correct, argued Norcross. The other two Nevada Supreme Court Justices, Patrick McCarran and Benjamin W. Coleman, agreed with the affirmation of judgment.[128]

Visual representations also played an important part in the decision of the Nevada Supreme Court in the case. All the models and maps that had been used in the original trial were shipped to Carson City for the inspection of the judges. The case itself was so dependent on visual representations that Norcross included three diagrams, printed in the reported version of the case, to justify his reasoning.[129]

Still unsatisfied with the results, the Jim Butler team appealed its case against the West End a final time, to the US Supreme Court. On March 26 and 27, 1918, Lindley argued the case before the US Supreme Court for the Butler, and Dickson for the West End.[130] Both sides largely repeated the assertions they had made to the Nevada Supreme Court. The Butler sought to have the lower court decision reversed because it violated the federal mining laws in three ways: that the end lines of the West End were not parallel and straight, that extralateral rights flowed in only one direction from the discovery, and that the facts of the case did not prove that the apex of the vein is in West End territory. The West End, in flowery language, argued for a liberal construction, based solely on the statute, unhampered by the precedents of the Leadville cases or others.[131]

Willis Van Devanter authored the US Supreme Court's opinion. Van Devanter often took a leading role in the Court's work on mining law and natural resource law generally, due to his comfort with and interest in the topic that stemmed from his experience in the West. He first addressed the question of the end lines, describing the context and using similar reasoning to that of Norcross to find that the West End's lines were valid.[132] Next, he took on the question of extralateral rights in multiple directions. Quoting at length from the 1872 mining law, Van Devanter carefully set out every way in which the law limited the rights of claimants. None of these included a single direction only, and thus, concluded the judge, no such restriction was intended.[133]

To resolve the third question, about whether there was a proven apex inside the West End claim, Van Devanter simply turned to the exact wording of the Nevada Supreme Court's summary of Averill's description of the vein and its surrounding ground.[134] Without challenging any of the conclusions of the earlier finding, Van Devanter repeated several of the salient points of the large passage he quoted and noted, "In these circumstances we hardly would be warranted in saying as a matter of law that the vein has no top or apex within the claim in the sense of the statute."[135]

Van Devanter concluded his opinion by suggesting that the West End's case was probably good enough. He reminded the court that it had earlier ruled that it was not right to take property from a discoverer and give it to someone with an adjoining mine that had been influenced to locate there because of the other miner's discovery. He offered no citation for this point, but worse it seemed to reverse the essential point that the Butler was asking for the removal of the West End from the Butler property, not the transfer of the veins in the West End claim to the Butler itself. Van Devanter noted that, in this case, no one claimed that the apex was somewhere other than West End ground, but only that what the West End had was not actually an apex. Without mentioning the distinguished lawyer by name, the judge served up a quote from *Lindley on Mines* to justify not overturning the ruling and giving the victory to the West End: "'The law,' as has been truly said, 'assumed that the lode has a top, or apex, and provides for the acquisition of title by location upon this apex.' Probably this assumption could not be indulged where the fact appeared to be otherwise, but it serves to show that the absence of a top or apex ought not to be adjudged in the presence of such a finding as we have here."[136]

As a result, the final appeal to the US Supreme Court preserved intact the most fundamental questions of the case, about the appropriate structure of the vein. Those geological questions were not seriously addressed beyond the initial trial, since later courts left Averill's findings of "fact" intact. Those findings, however, were based in significant part on the visual representations, especially the three-dimensional skeleton model of the West End, presented as evidence in the trial. The visual representations were not evidence, however—they were finely crafted arguments about geology, and the "facts" of scale, size, shape, location, and type became strongly and almost indetectably intertwined with arguments about the space in dispute.

The *Jim Butler* decision had a relatively minor legal aftermath as a precedent for other cases. The period around World War I saw the last major developments in apex litigation, of which the *Jim Butler* case was a part. After that time, there were relatively few apex cases tried. There were several reasons for this change—the increasing cost of such litigation, well borne out by the Tonopah example, was one, but a larger reason was that the mining industry was changing in such a way as to render obsolete most apex litigation. The increasingly high capital costs of mining necessitated development on a larger scale than before, which tended to mean that a company bought all the claims nearby before commencing major operations. Additionally, the postwar emphasis on larger, lower-grade deposits (such as the porphyry copper lodes of Arizona, Utah, and eastern Nevada) instead of traditional mineralized fissure veins tended to reduce fights over extralateral rights, since the huge low-grade deposits could hardly be said to have apexes at all.[137] Finally, few new cases were brought after World War I because most of the legal questions pertaining to extralateral rights were settled well enough to allow lower courts to manage the questions themselves.[138]

The *Jim Butler* case, though cited a handful of times in lower courts, was cited explicitly in a US Supreme Court case only once. This was the 1921 trial of *Silver King Coalition Mines Co.* v. *Conkling Mining Co.*[139] W. H. Dickson and A. C. Ellis Jr., and perhaps even Lindley himself, were lawyers for the defendant.[140] The lawyers for the defendant cited the *Jim Butler* case to "definitively and positively" show that mining companies have as many rights on secondary veins as they do on the discovery vein.[141] US Supreme Court Justice Holmes wrote the opinion, and cited the *Jim Butler* decision twice. The first time he used it, along with other precedents, to support the fact that if a vein strikes across a claim's side lines (instead of the end lines, as is supposed to be the case), then the side lines become the end lines and vice versa.[142] The second, more important, use of the *Jim Butler* case was to support Holmes' refusal to revisit the question of the specific geology of the vein (or veins). "We have the distinct testimony of experts that there was no such [other vein] and we agree with the view of the District Judge sustaining the petitioner's extralateral rights. Whether there are other answers to the contention we need not decide."[143]

Perhaps ironically, then, the most lasting judicial implication of the *Jim Butler* v. *West End* case was the decision that if the district court gets the geological facts right, higher courts need not consider revisiting them. Thus, three-dimensional models and other visual representations, which featured arguments about geology and law in a more fact-like form, were seen to have improved the decision-making ability of the lower courts, at least in terms of the facts of mining law disputes. The rhetorical power of the visual representations contributed to limiting the review of the information they purported to contain. To use Judge Averill's term, the information presented by the models was so "obvious" that the highest court in the land could feel free to safely ignore it as settled.

The outcome of the trial meant ruin for the Jim Butler and prosperity for the West End. The stock of both companies had increased in value during the suit, because the West End was allowed to continue to mine and mill the ore. All of the profits from the ore were placed in an escrow account, which totaled more than $400,000 by the time the last appeal was finished. This figure represented profit from the ore only, and did not include the triple damages that were at stake. When the Butler lost the suit, the West End gained control of the only productive parts of the Jim Butler holdings. The Jim Butler was reorganized to pay down some of the debt of the Tonopah Belmont as well. Lacking proven reserves or the capital to conduct a full-scale exploration for more ore, remaining pockets of the Jim Butler were leased to small miners for more than a decade, but income from this source was minuscule, and the company never regained anything like its pretrial prosperity. In 1938 the Tonopah Mining Company bought control of the Butler for less than $3,000 and conducted some exploratory drilling, but nothing worthwhile was found, though leasers occasionally shipped out ore as late as 1947.[144]

The trial had a significantly better outcome for the West End Consolidated Mining Company, as might be expected. The company made significant profits in the years during and immediately after the trial (except for 1919, when a large strike hurt production at all Tonopah mines). Prosperity led to investments in mines outside of Tonopah, which proved to be a drain on the company: profits from the West End's Tonopah property financed the costly experiments elsewhere. Even with the baggage of failing mines, the West End continued to make profits, though on a diminishing scale, into the mid-1920s, largely on the strength of its Tonopah output. (Radical drops in the price of silver in 1923 also significantly affected the profitability of the West End and other Tonopah operations.)[145] The trial itself was expensive—it cost the company almost $115,000—though the victory put the West End firmly in the black. The company spent $79,469 on trial preparations such as development work and model making, and lawyers' fees for the trial amounted to an additional $35,000.[146]

The little cabal of mining litigation experts were doubtlessly the least affected by the outcome. For Lindley and Dickson, the distinguished old lawyers, the decision came in the twilight of their careers, and neither would live much longer. Younger lawyers and experts, including Finch, Searls, and Colby, had long careers and significant accolades in their futures. Some of the locals, especially Brown, Atkinson, and Judge Averill himself, continued their careers in Nevada and tackled questions of mining law only infrequently thereafter.

After the trial at the US Supreme Court, all of the exhibits were returned to the companies. Francis M. "Borax" Smith, president of the West End, donated the skeleton model to the University of Nevada's Mackay School of Mines shortly after the conclusion of the trial; this model, along with mineral specimens, served as silent teaching tool in the school's museum. In 1939 the West End asked to borrow the model back from the University, as the company hoped it "would be of real economic service to us for a time." It is not clear if the loan was made, but the model clearly returned to the museum again, where visitors can see it today—a perplexing and captivating representation of a now-invisible underground landscape.[147]

Technical models, such as the West End model now on display at the Keck Museum, were extraordinary new ways to peer into the earth. The models in this case carried a large amount of power. Not only did both legal teams use the models to help the judge visualize the space of the mine, but also the judge used one of the models in his decision-making process. Visualizing the underground space was paramount to helping the court understand the issues in the case. And, for better or worse, the models presented by both legal teams gave the judge the information he needed to render his verdict.

As noted in chapter 5, models of underground spaces broadly belonged to one of three typologies: block models, sectional models, and negative or skeleton models. The characteristics of each type of model suited them for different types of representational work. Block models showed geology well, but did not depict underground excavations easily. Sectional models made of glass were inexpensive and easy to make, but were not truly three-dimensional. Negative or skeleton models were fragile and looked strange, but showed mine workings in three true dimensions. The choice of a model type was the choice of a limited range of representational types, as the legal teams of the Jim Butler and the West End certainly knew.

Representational styles could directly affect the effectiveness of a model as a form of technical communication. This point is brought out clearly by the *Jim Butler* v. *West End* case. The visual representations used by each side, especially the three-dimensional models we focused on here, helped shape specific understandings of those underground landscapes that conformed to the demands of federal mining law. Both sides embodied arguments about underground spaces in the models and maps they made. These representations, though, transformed *arguments* into *facts*—facts that were then used to support arguments about geology and the law. The West End's coherent and consistent system of interlocking visual representations, anchored by the three-dimensional model, performed this rhetorical sleight-of-hand most effectively. By contrast, the less consistent and occasionally contradictory messages offered by the Butler models undermined their trustworthiness in the eyes of the court.

In the trial, the visual culture of mining at its most advanced was at the center of a complex legal problem concerning the properties of an invisible underground space. Below the surface of the earth was rock of different types, spaces carved out by miners, and perhaps some machinery, but the visual culture of mining transformed it into a geolegal subterranean landscape, accessible only to experts and lawyers, ruled by abstract concepts, and divisible with a judge's gavel.

chapter 6

Mine Models for
Education and the Public

The visual power of mine models made them perfect vehicles for teaching. Such models helped mining engineers convey, in a visual way, the spaces and technologies of industrial mining that they helped to create. Although mine models varied widely in their content and form, all helped engineers reach new audiences.

One audience for a simplified understanding of industrial mines were young mining-engineers-to-be. Models could help students who had not yet developed a knack for making three-dimensional pictures in their mind's eye visualize the complex spaces of a modern mine. Given the difficulty of seeing many mining technologies at full scale underground, a model could represent all at once a complete system or make a technology visible that could not be seen in the darkness below.

Models could transport a mine to visitors, in order to reach audiences that could not view the site itself. Beginning in the late nineteenth century, mine models became increasingly popular at the great world's fairs. The public embraced the opportunities for entertainment and education the fairs provided. Corporations grew to realize the immense opportunities to showcase their goods and services. In this context, models of mines made by corporations served commercial as well as educational purposes. After the fairs, such models often ended up in museums, to continue portraying their educational (and corporate) messages. With the help of like-minded curators, these models brought the ethos of mining engineers and the corporations for which they worked to the exhibit halls of American museums. One well-documented example is the mine models displayed in the United States National Museum, part of the Smithsonian Institution, in the early decades of the twentieth century.

Early technical models for education, especially those specifically representing mining engineering problems rather than more general geological questions, were generally made overseas. As mentioned in chapter 4, J. P. Lesley and several members of the Harden family made geological and block models in eastern Pennsylvania beginning in the 1860s, but their preferred materials made it difficult to make duplicates for mining schools located beyond the Eastern seaboard. The Hardens repeatedly discussed the problem of making transportable models. Harden normally made plaster negatives of his wooden and wax topographic models, so he could make plaster positives, which lent themselves to customizable coloring and duplication. The trouble was that plaster was expensive, heavy, and fragile. Harden and Harden moped, "[I]t is desirable that a cheaper method of duplicating models be found. . . . Plaster of Paris is for many reasons a good material; but its weight and its liability to damage in transportation render it unsuited."[1] The Hardens conducted experiments for years hoping to find a method of using paper or some other lightweight and inexpensive material to reproduce the topographical models, but never reported the success they sought.

Overseas mining schools developed models and similar methods of instruction before they became widespread in American institutions. William Jones described for a popular audience the mining school of St. Petersburg, in the Russian Empire, and its attendant museum. In addition to mineral specimens and examples of tools and machinery, "There are also *fac-simile* representations, in miniature, of several mines, showing in what manner the metals are worked, and the kind of machinery employed."[2] More striking than the jewels or the models, however, was Jones' trip into the full-size training mine, attached to the museum and used by students: "A guide precedes the visitor with two candles in hand, and after passing through several passages, and unlocking two ponderous doors, he is led into the gloomy recesses of a *counterfeit* mine, absolutely underground, furnished with all the machinery, tools, and implements used in a real mine. Here the pupils of the mining college are instructed, by an exact representation of various mines, how the different ores are found."[3]

Freiberg, in Saxony (Germany), was well known as a center of mine model making as early as the eighteenth century, and the mining academy there made extensive collections of models to further their students' education. J. C. Bartlett reviewed the attractiveness of the foreign mining schools to prospective American mining engineers in the mid-1870s. Bartlett held highest regard for the mining academy at Freiberg, which had a longstanding reputation in mining engineering education. In addition to exposure to some of the top professors in the field and a wide range of courses, Bartlett also pointed out the collections of study aids: "Moreover, the large and valuable collections of minerals, geological specimens, and models of machinery are accessible for inspection or study."[4] Though Freiberg's museum of mine models gradually became less reflective of current mining practice in the early twentieth century, models continued to be made in Freiberg, and other German universities made extensive collections of up-to-date material.

Some Freiberg models eventually made their way to American colleges, but the demand for mine models exceeded the supply. According to H. H. Stoek, professor of mining engineering at the University of Illinois, quite a few of the models found in America in 1917 came from Freiberg, and his own institution owned a Freiberg model that was made before 1892.[5] Walter R. Crane, professor of mining engineering at Pennsylvania State University, reported that he had been interested in making mining models since the late 1890s, and other mining professors reported some attempts at modeling.[6] Even so, Stoek lamented that the use of models in mining education was underdeveloped in America, "chiefly because of the cost of the models and the scarcity of model makers."[7] Professor F. W. Sperr, of the Michigan College of Mines, pointed out that although he had been making models occasionally for many years, "I felt the need of them for a long time before I began to make them."[8]

In some cases, mine models made for lawsuits or other purposes were donated to mining schools. For example, the Montana School of Mines (today's Montana Tech) received many litigation models from Anaconda after the lawsuits that inspired their creation were concluded. (A handful of these models are still on display.) Similarly, the company donated the model of the West End Consolidated made for the mine's lawsuit against the Jim Butler (described in chapter 5) to the Mackay School of Mines at the University of Nevada. However, such donations for educational purposes were, according to one model maker, "rare and fortunate," because the models usually contained so much information about the underground workings considered by the companies to be "of a private character" that the models were locked away in storage or, worse, destroyed at the conclusion of a lawsuit.[9]

While some models in use in universities came from mining companies after having been made for lawsuits (such as the West End model described in chapter 5), in other cases mining companies donated realistic models that had been meant for commercial purposes. For example, in 1895 the Russell Process Company, who employed their eponymous patented lixiviation milling technique at the Marsac Mill in Park City, Utah, donated a large model of their plant to the Columbia College School of Mines. The company had built this model, thirteen and a half feet long and four and a half feet deep, exhibited it at the Chicago world's fair. The company planned to donate it to a mining school, and contacted Columbia first with the offer. Thomas Egleston reported that this model was the most valuable donation ever received by the metallurgy department at Columbia, and he was happy to have it despite needing to spend several hundred dollars to build a glass case to protect it.[10] The St. Joseph Lead Company of Missouri also intended to donate a model the company made for the 1893 fair to Columbia's School of Mines, though, according to company officials, many of the other Missouri mining displays were slated to go to that state's public school of mining in Rolla. Thus the major exhibitions helped build teaching collections, despite the commercial origins of the models.[11]

More commonly, instructors of mining engineering took the matter into their own hands, building models for use in their classrooms. Stoek described a series of four models that he built in the 1910s to help him teach mining engineering as it was practiced in horizontal coal beds like those of Illinois. Figure 6.1 shows one of them. His models were large and flat with little relief, made of wood and surfaced with coal dust to achieve a realistic look. He mounted them in wooden frames that traveled on casters and could be rotated on the horizontal axis, like old-fashioned movable blackboards, so he could both position them optimally for viewing and also fit them through the classroom doors.[12]

Walter R. Crane faced similar issues in finding models that were suitable for use in his classroom. Models were of "unquestionable" value to teach mining engineering, noted Crane, but the two main purposes for which

Figure 6.1. Henry Harkness Stoek, who taught mining engineering at the University of Illinois, designed this and similar models to demonstrate to his students various methods of coal mining. The model was large and flat, mounted to a frame like a blackboard, and covered with glued-on coal dust to give the correct appearance. From H. H. Stoek, "Mine Models," *Transactions of the American Institute of Mining Engineers* 58 (1918): 27.

models had been made both worked against their usefulness in classrooms. Models made for display in expositions and museums were too "superficial" and did not contain information precise enough to be useful to mining students.[13] By contrast, models made by mining companies were accurate but ugly, and sometimes difficult to understand. For educational use, Crane experimented with making realistic mine models out of wood and concrete, where concrete's ease of use while wet and durability once dry were of significant importance to the modeler. (His concrete version of a gold and silver mine can be seen in figure 6.2.) While Crane's painstaking efforts yielded impressive models that represented idealized practices for his mining students, their creation, as he described it, was likely so tedious as to discourage other would-be modelers from following his example.[14] Crane used photographs of his models as illustrations for the second edition of his textbook on mining methods, and for a separate article.[15] Crane noted that models were intentionally idealized representations, intended to foreground important principles and deemphasize confusing or irrelevant detail. Crane hoped to provide his audience and his students with a "comprehensive grasp" of different methods of mining, enabling his students to "see . . . the mine in its entirety."[16]

F. W. Sperr, professor at the Michigan College of Mines, tried a wide variety of materials and methods in making models for his classroom. Sperr noted that building educational models required making decisions, and often trade-offs, among matters of realism, comprehensiveness, material workability and cost, and what should actually be shown on the model. He tried making models out of actual rock, but (perhaps predictably) the results were too heavy and not comprehensive enough. He tried a series of wooden models, chosen to illustrate the different stages of mining, but it was difficult to make wood realistically look like the rock of

Figure 6.2. Walter R. Crane made this realistic model of a stope from concrete. He saw models like this one as important educational compromises between more-technical models that could not be understood without training, and inaccurate models that did not contain enough information to be instructional. From W[alter] R. Crane, "Concrete Mine Models," *Mines and Minerals* 27 (February 1907): 301.

an underground mine. Sperr also experimented with papier maché, paste-board, and wire mesh covered with a magnesia coating, but none of these completely satisfied the professor.[17]

While there were relatively few model makers creating mine models for engineering classrooms in the early twentieth century, at least one com-pany making mine models bridged the gap between technical models for lawsuits and those for classroom instruction. Frank A. Linforth and E. B. Milburn were geologists for the Anaconda Copper Mining Company in Butte. (Chapter 3 discusses the pioneering work of the Anaconda's geo-logical department.) Apparently, these geologists had a knack for making models. They set up the Engineering Model Works of Butte, with head-quarters in the same building as the Anaconda geological offices. As the Engineering Model Works, they built several models for Anaconda's use in mining lawsuits, mostly of the negative or skeleton type. However, Lin-forth and Milburn also made technical models for mining engineering classrooms. Most intricate were their scale models of mining headframes, made of heavy drawing paper very precisely cut, folded, and glued (figure 6.3). Their paper headframes, which stood three feet high or more, were light, strong, and inexpensive enough for engineering programs to afford; the firm reported many sales to colleges and universities. The duo also used similar techniques to model other mine equipment, from mine cars, skips, and bridges to complex smelters and concentrating plants.[18]

Mining professors acknowledged the importance for mining engineer-ing students not only to learn how a mine worked, but also to see its inner workings and the relationships of engineering systems underground. Crane believed that the opportunity presented by models to see mines all at once and in three dimensions improved the work of his students. Because of the new way of seeing permitted by the models, Crane noted, "I can confi-dently say that the grade of work done in the classroom has improved in a surprising manner."[19] However, the pedagogical benefit of models was not automatic. F. W. Sperr, who taught at the Michigan School of Mines, found that simply having the student look at the models as a professor described them "failed to increase the efficiency of our teaching as much as we had anticipated that it would do." Sperr also anticipated it would be redundant to have models of mines that the students could visit in real life. To his surprise, the students learned particularly well when models were com-bined with visits. He had his students sketch and describe the models prior

Figure 6.3. This model of the New Leonard headframe was created out of heavy paper, cut precisely to size and glued together, by the Engineering Model Works of Butte, Montana. That firm, made up of geologists from the Anaconda Company, mostly made models in support of the mining firm's extensive litigation, but also created models such as these for pedagogical purposes. From "Head-Frame Models Made of Paper," *Mines and Minerals* 30 (February 1910): 401.

to visits to their underground analogue. Using their notes from the models as a reference, the students were able to grasp underground operations much more completely and quickly than before.[20]

The sheer size and awkwardness of models could make it difficult to use them effectively in educating engineers. At the Columbia College School of Mines, Professor Thomas Egleston used models to help teach his students about metallurgy. These models were stored along with the other mineral specimens in a small museum room near his lecture hall. This meant that students had to visit the models in groups, because the "expensive and very heavy" models could not be easily moved and the specimen room could not accommodate everyone at once. As a result, Egleston would show diagrams and describe the features depicted by the models, but the students would have to visit them later. In 1890 declining enrollments prompted Egleston to request the removal of a row of seats from the lecture hall and the installation of the models in new cases along the wall, so they might be visible as he lectured.[21]

Models could also be used in a different pedagogical fashion, to educate those students that might be labeled "tactile learners" in a more modern age, through learning about mines by constructing models. Bennett Brough reported in 1888 that students of the Stockholm School of Mines in Sweden were required to learn how to build glass plate mine models.[22] Joseph Daniels, who taught mining engineering at the University of Washington, noted in 1917, "[T]he greatest good in the development of the model comes to the person who makes it." At his institution, freshman mining engineers were required to build models of shaft timbering, and Daniels mused that it might be a good idea to have every mining engineering student make a model sometime during his collegiate education.[23] By requiring students to individually build a mine model, instructors implicitly forced them to understand a mine and to express their understanding in a visual way.

Educational models might also have less predictable effects on young engineering students, such as serving simply as inspiration. Metallurgical engineer Edward P. Mathewson, who won worldwide fame as head of Anaconda's copper smelters, stumbled on his future career by accident. The son of a Canadian grocer received the first spark of interesting in mining and metallurgy in the early 1880s at McGill University: "My first impulse toward metallurgy came when looking at some models of furnaces in the college at McGill. When I entered McGill I did not know what course I was to follow, and went through the first year without making a decision."[24]

No matter what their origin or form, mine models used in the education of future mining engineers represented a use of the visual culture of mining directed toward those who would eventually grow to incorporate it in their professional lives. Beginning in the late nineteenth century, mine models were also mobilized to educate audiences beyond those of mining engineers—the general public itself. These public-facing exhibits, in expositions and museums, can be understood as an expression or reinterpretation of the visual culture of mining, supporting the activities and values of mining engineers. Even though these exhibits were far from the mines, the models they showcased were fruits of the ideology of mining engineers, and the visual culture of mining was used to achieve the engineers' aims.

Mine models were rare in American museums and expositions before the 1890s. Mining as a subject certainly received attention, but largely took the form of individual specimens of minerals, ores, and products. For example, the early collections of the Smithsonian Institution contained some geological specimens. The Centennial Exhibition in Philadelphia in 1876 resulted in a flood of new material to the Smithsonian, and prompted the formation of the United States National Museum as a branch of that institution with a building funded by Congress.[25] The centennial collections contained mining materials, but as before most were small specimens of rocks, ores, and finished materials. The United States National Museum later received additional mining materials from the 1876 exhibition, that had been originally donated to the AIME at the close of the Fair. These too were specimens of minerals, ores, and products. Frederic P. Dewey, an active member of the AIME who had worked in the minerals industry, was hired as a curator in 1881 and worked to build and display the collections. He solicited donations directly from the mining industry and sent young mining engineers to seek out specimens on his behalf, with an eye toward building displays for the 1884 New Orleans exposition that could later be installed permanently in the United States National Museum in Washington. This exposition, like the earlier 1876 gathering, also gave Dewey an opportunity to secure donations of mining-related material from other exhibitors.[26]

Dewey remained engaged with the AIME throughout his time as a curator. In fact, he combined trips to the AIME meetings with collecting expeditions, obtaining specimens and promises of additional material both during the multiday meetings and during extended collecting trips afterward. He also continued to present and publish professional papers in the *Transactions* of the AIME and other journals. Some of his pieces were short, consisting of detailed analyses of particular ores or specimens, much like his published efforts before his arrival at the museum. Others were research efforts based on interesting collections that belonged to the museum, and still others explained and promoted the museum's work to his audience of mining engineers.

The mining items on display in the United States National Museum in the 1880s and 1890s overwhelmingly consisted of rocks, ores, and bits of metal, along with some photographs, but Dewey did have at least four models in his collections. In 1884 the museum purchased a copy of Charles Ashburner's topographic model of the Panther Creek coal basin, which had been made using underground contours. The model was displayed at the New Orleans Exposition in 1884, and returned with Dewey for permanent installation in Washington.[27] The next model Dewey received was of the "Ruby Hill mines," this was likely one of the models used in the *Eureka-Richmond* case in the late 1870s, described in chapter 4. (Both mines were situated on Ruby Hill.) This complex model was in the hands of the museum by July 1886, when Dewey's assistant reported spending several days struggling to set it up.[28] In 1888 Dewey received a glass model of the New Almaden mercury mines in California along with specimens and other materials, as the result of an offer from James B. Randol, in charge of the Quicksilver Mining Company.[29] Shortly thereafter, the museum was given a block model of the mines of Aspen, Colorado, that had been used in a lawsuit, and that likely was the model made for David W. Brunton and described by him in the *Engineering and Mining Journal*.[30] The New Almaden and Panther Creek models were described in the catalog of the systematic collections, implying they were on display at the time Dewey departed in late 1889. The Ruby Hill model was apparently set up in 1886, and though it was not described in Dewey's report of the collections at the time of his departure, it was on display in 1892.[31] However, it is not known if the Aspen model was ever put on display. In any case, in the United States National Museum through the 1880s, the models were few, largely reflecting exhibiting practice at world's fairs in that period, and half of those few models museum did have had been originally created for lawsuits, not promotion.

Models became more common at world's fairs beginning with the great Columbian Exposition of 1893 in Chicago. This Chicago World's Fair featured an entire building devoted to mining, which was subdivided into exhibits of individual states and foreign countries, as well as displays by mining companies and equipment manufacturers. Many of the state exhibits featured a model or two, typically donated by a deep-pocketed mining corporation.[32] As mentioned above, some of these models made their way to educational institutions at the close of the Fair. However educational the exhibits appeared to have been, they could also be entertaining to their visitors. One elaborate model, the "Colorado Gold Mine," illustrates this well. Constructed by William Keast to depict the inside of a mine, this gigantic model was located not in the mines building, but on the midway with the other entertainment. It was so large that visitors walked through it—provided they first paid a dime's admission.[33]

While the United States National Museum received no mining materials from the Columbian Exposition, in part due to a lack of space to receive them, future expositions would yield mine models for display in the museum. One reason for the change was the availability of new space. The National Museum had pushed for a new building to house its ever-expanding collections within just a few years after the first museum building opened in 1881, but like all things in Washington, this moved slowly. The ever-increasing space squeeze was part of the reason Dewey's metallurgical displays were gradually removed from view in the 1890s. Congress was persuaded in the early twentieth century of the need for a new museum building to better protect and display the collections, and the museum broke ground in 1904 on what would become the National Museum of Natural History building. Though it took years to complete—a handful of collections moved in 1909, but the new building was not fully occupied until 1912—the immediate prospect of additional space permitted museum officials to plan for new areas of inquiry in the expanded museum.

Before the new building was even complete, an opportunity to collect artifacts about mineral technologies presented itself to the museum in the form of the Louisiana Purchase Exposition of 1904.[34] Administrators wanted to develop more mining exhibits in the museum once the new space was available, and the 1904 Exposition, held in St. Louis, contained a large number of exhibits on mineral technology. Administrators tapped Charles D. Walcott, head of the USGS and honorary curator of stratigraphic paleontology at the museum, to collect and store materials for the future collections, and gave the curator wider latitude than even the other Smithsonian employees collecting in St. Louis.[35] Walcott ended up with thirty-five railroad carloads worth of stuff—not only American artifacts, but also materials from Austria, Belgium, Brazil, Canada, Cuba, France, Germany, Great Britain, Italy, Japan, Mexico, Peru, the Philippines, Portugal, and Siam (now Thailand).[36]

A lack of space meant nothing was done with the exposition materials for years. Walcott had become secretary of the Smithsonian by the time exhibits could be moved to the new building, freeing space in the old one to set up the mining exhibits.[37] In 1913 efforts to create mining exhibits in the museum finally got under way. Chester Garfield Gilbert was the curator charged with developing the division of mineral technology. Gilbert came to the museum in 1911, a 1905 graduate of the University of Rochester who briefly taught at Lehigh University. He cut his teeth at the museum working to separate any mining-related items, left over from Dewey's displays of the 1880s, from the geological collections.[38] As Gilbert began to unpack the material from St. Louis, he decided, "[T]he greater portion of the specimens proved to be wholly unsuitable for use along the accepted lines of development of the division, and were either returned to the donors or destroyed."[39]

Though today the Smithsonian museums emphasize historical items, this was definitely not the case in Gilbert's day. Instead, the curator's plan was to actively assist the mining profession by displaying up-to-date materials. He clearly had little place in his plans for obsolete technology suitable only for historical displays. For example, Gilbert moved to discard a stamp mill from the 1850s, which was originally part of the display of the State of California at the 1904 exposition, as it had "no value excepting from an historical view point to the State of California."[40] A museum interested in the history of American mining would likely view a California stamp mill from the 1850s as a crucial artifact that illustrates the development of American mining technology. Gilbert simply did not believe that history of mining technology was relevant to furthering his promotional aims. Gilbert did retain some of the St. Louis materials that were up to date, including several mine models and some large photographs of a modern coal mine suitable for display.

Gilbert declared that his division would first serve as a kind of information clearinghouse, connecting the results of scientific research performed by industry to the general public, and erasing any doubts as to their efficacy along the way. The curator believed that information directly from companies risked rejection as mere advertising by the public. Gilbert noted that new information about the properties and uses of minerals was being created on a regular basis, however:

> For the dissemination of this mass of most important information the public is almost wholly dependent on the industrial advertising manager, and however accurate may be the contributions from such sources, they are bound to fail in their broader educational value through the fact that the information does not emanate from a disinterested source. In its most purely technical aspect, therefore, the real opportunity of the division to be of service lies, not in the direction of abstract research, but in the exactly opposite one of rendering assistance toward keeping the public in touch with important current developments in mineral technology.[41]

Perhaps even more than other divisions in the museum at the time, Gilbert's Division of Mineral Technology relied on three-dimensional models to convey the curator's arguments to the visiting public. Some of the models, particularly those created in the late 1910s and early 1920s, were made by in-house modelers, who usually worked from plans for the authentic artifact, supplied by the companies being portrayed. Other models, especially those placed on display in the early years of the Division of Mineral Technology, were created by the companies and acquired for the United States National Museum, often after they had been displayed at exhibitions. Lifelike display models of many sectors of the mining industry were eventually put on display, but below we examine closely two models of coal mines.

Gilbert had not been responsible for building or collecting the models from the 1904 exposition, but he readily embraced the pedagogical potential of such exhibits. Though he had inherited the first models as a consequence of the collecting activities of Walcott, Gilbert soon envisioned a "series of models designed to show important variations and adaptations in mining procedure" for his exhibits. He wrote to the deans of several prominent mining engineering programs for advice about his plan. Gilbert argued for the importance of models in his displays: "In studying unfamiliar drawings about 75% of the average person's capacity for concentration goes into the effort to project his imagination into a third dimension. Models eliminate this tax on the imagination and at the same time facilitate attention."[42] Similar to mining lawyers' and educators' understanding, Gilbert clearly if implicitly understood the importance of models for communicating mining engineering ideas to nontechnical audiences. An engineer generally trained his mind's eye, over time, to readily project two-dimensional drawings in three dimensions, but the general public did not. Thus, models were important to Gilbert because they could convey information to his targeted audience that would have been lost in two-dimensional displays in the older tradition.

As Gilbert explained his scheme in his letter, his initial idea was to have plaster models made that illustrated particular features of mining processes, then display the models at the museum and circulate photographs of the models in order to spread their pedagogical reach. He wanted to see if the Columbia dean, who was well connected to the mining industry, believed that the educational possibilities of such a scheme might justify the expense. Gilbert received a reply not from Dean Goetz, but from Rossiter W. Raymond, one of the greatest names in American mining engineering. Raymond, in the twilight of his career in 1916, served as an elder statesman for the mining industry and was closely connected with the Columbia College School of Mines; Dean Goetz had asked Raymond personally to reply to Gilbert's letter. Raymond supported Gilbert's use of models, noting "I believe quite strongly in the utility of mine models for illustration to students, and for exhibition in museums and institutions," though he demurred on whether they would be sufficiently valuable to justify making a large number of them for Gilbert's Smithsonian patrons. Raymond then offered Gilbert some practical advice about model making, suggesting some sort of lightweight stucco instead of Gilbert's proposed plaster, and noting the utility of glass plate models. He closed his letter to Gilbert by remarking that he was "very much interested" in Gilbert's project and would be happy to help in any way possible.[43] This correspondence suggests the extent to which the use of models in certain realms of mining engineering practice, as we outlined in chapter 4, had started to inform museum practice. Ultimately, however, Gilbert's model displays were influenced, especially at first, by the materials from the 1904 exposition that were already in his curatorial custody.

The first artifact entered in the catalog book by the Division of Mineral Technology on its organization in 1913 was a very large model of the New England Mine of the Fairmont Coal Company, located in Fairmont, West Virginia (shown in figure 6.4).[44] The model was constructed at a scale of one inch to one foot, so as to appear impressively lifelike. The machinery had been modeled to perfect scale, "even down to the bolts and rivets," and steel-wool smoke issued from the plant's stacks. The miner's village portrayed on the model depicted a host of tiny details including telephone poles, lawns, and flowers, "in fact, everything to make the picture a natural one."[45] The model's popularity was helped by its substantial size (thirty feet by forty feet), as well as by its height eighteen inches off the ground: this permitted children a view and adults the opportunity to see details from above. The real attraction, however, was the electric mine locomotive that moved in and out of mine tunnels, feeding the machinery that moved coal back and forth. Very few exhibits in the United States National Museum in 1914 moved at all, so such a display was quite popular. Rather than leave the model running all the time, which would require a continuous staff presence to ensure nothing went wrong, the museum put up a sign by the display informing visitors that the model would be activated every hour. The museum guards were supposed to show up and operate the model, but on several occasions curator Gilbert showed up ten or twenty minutes after the show was scheduled to find the gathered, waiting crowd disappointed because the watchman had failed to arrive.[46]

The museum claimed that the Fairmont model would allow "the visitor's imagination [to] visualize accurately the social conditions typical of a coal-mining community."[47] The model, for all of its rivet-accurate verisimilitude, would have been misleading at best in the portrayal of the "social conditions" of coal mining life. The model houses were clean and their closeness to the mine would have suggested only a conveniently short commute, instead of their actual proximity to loud and dirty coal operations. The dwellings on the model looked sturdy, and each had steel-wool smoke issuing from its chimney. This fake smoke looked like a smaller version of that spewing from the stacks of the coal plant, and thus drew an implicit comparison between the seemingly orderly, rational, clean industrial operation and its apparent happy domestic counterpart, likely presided over by the virtuous and industrious miner's wife. The model could not, or would not, speak to labor trouble or the dangerous work of mining. This is especially ironic because the worst mining accident in U.S. history, the December 1907 explosion at Monongah, West Virginia, that killed 362 miners, occurred in a similar coal village owned and operated by the same

Figure 6.4. This model of the New England Mine of the Fairmont Coal Company in West Virginia had been originally constructed by the company for display at the 1904 St. Louis Exposition. At the close of the fair, it was donated to the United States National Museum, part of the Smithsonian Institution. It was put on display there in 1913, and the model, as shown in the photograph, was large and comprehensive. It remained on display until World War II. Image MAH-28515, Smithsonian Institution Archives.

company. The horror of the Monongah explosion even provided the impetus for the formation of the U.S. Bureau of Mines, because the mine operators seemed so clearly incapable of preventing such disasters.[48]

The model's pro-company slant is of little surprise, of course, because it had originally been constructed by the Fairmont Coal Co. for the St. Louis Exposition of 1904. There, as seen in figure 6.5, the Fairmont model was only part of the large display of the Consolidation Coal Company and its semi-independent subsidiaries such as Fairmont. Two more models of company operations in Maryland and Pennsylvania shared the company's pavilion space. The visitor was expected to be able to contrast the different mining methods used by the companies. All were up to date and state of the art, but their differences showed the exposition visitor the expertise of Consolidation in adapting their methods to suit the particular geology of each region. All three models showed similar scenes of clean, tranquil towns, suggesting the company's benevolence, but the juxtaposition of the three models focused the visitor on the models' technical content. Once the Fairmont model was detached from its exposition context and placed, alone, in the center of one of the halls of the United States National Museum, the new setting downplayed the model's portrayal of up-to-date coal mining methods, and instead naturalized the clean and efficient company town as a primary point of the display. This model portrayed its subtle messages to museum visitors from its installation in 1914 until it was finally dismantled and scrapped in 1943—nearly four decades after it had first been put on public view at the exposition of 1904.[49]

A model of the First Pool No. 2 mine of the Pittsburgh Coal Company, seen in its display case at the United States National Museum in figure 6.6, provided another potentially misleading portrayal of the mining industry.[50] Like the Fairmont coal model, the Pittsburgh coal display was originally constructed for the company's exhibit at the St. Louis World's Fair of 1904, and was collected by Walcott for the museum at the close of the exposition.[51] The Pittsburgh model also was long-lived, and was not scrapped by the museum until 1960. Like the Fairmont model, the Pittsburgh model was touted for its truthfulness, a 1:48 scale three-dimensional

Figure 6.5. The large model of the Fairmont Coal Company at the United States National Museum was originally created by the firm for display at the 1904 St. Louis World's Fair, seen here on the right. Its proximity to other commercial products at the exposition conveyed a different message from the one conveyed byits later display in the museum. From "Coal Mining at Louisiana Purchase Exposition: Description of Models Showing Works of Some of the Large Bituminous Coal Companies," *Mines and Minerals* 25 (September 1904): 83.

Figure 6.6. The model of the First Pool No. 2 Mine of the Pittsburgh Coal Company is seen here in its display case at the United States National Museum. This model depicted a wholesome miner's village as well as underground workings at a 1:48 scale. Scanned from a photograph in the Mineral Technology Divisional Records, Division of Work and Industry, National Museum of American History, Smithsonian Institution.

picture that "copies faithfully the surface conditions at the mine."[52] Unlike the larger Fairmont model, however, the Pittsburgh scene depicted the underground as well as the surface of the mine. The so-called truth of the model thus extended to its portrayal of underground conditions as well.

The Pittsburgh Coal company model showed a clean industrial operation sited just below a wholesome little village on a hill. The prominent place of the white church, alone by itself on the hill, showed that the miners were a god-fearing people, and the conspicuous white schoolhouse (largely obscured in figure 6.6; its belfry and roofline is just visible beyond the roof of the large tipple building in center right) showed the benevolence of the company in providing for its loyal workers and their families. The village in the model seems a picture of tranquil preindustrial small village life, despite the fact that some 255 men worked for the mine at the point in time that the model was supposed to represent—a full-bore industrial firm.

The peaceful and harmonious scene extended below ground. While a miner on his way to work can be seen just in front of the rightmost house and another supervises the mechanical loading of coal into a railroad car, most of the human figures visible in the Pittsburgh coal model are underground. There they work in what seems like a relatively clean, safe, and spacious system of hallways. Many of the common hazards of mining were due to falling rock and coal from the interior roof, which was not depicted on the model. The coal seam worked in the actual mine depicted by the model was only five and a half feet thick, so any miners of above-average height would have had to stoop throughout the day, though no figures are shown on the model in the characteristic crouch of low coal mines. No coal cutting machinery is visible either, though pneumatic coal-punchers were used in most parts of the mine.[53] The workers, all men, all have clearly ethnically white complexions (though some do sport handle-bar mustaches) and none exhibits the sort of coal grime that made coal miners instantly recognizable in the photographs of Lewis Hine. A more subtle suggestion of false harmony was evident in the underground layout. While most companies would work diligently to avoid digging for coal directly under their surface facilities and railroad tracks (undermining) for fear that the land would subside, the companies generally had no such compunction about mining under the nearby towns. In the Pittsburgh model, however, the opposite is portrayed—the tunnels extend under the coal tipple and railroad, but do not reach as far as the town. Thus the viewer of the model might detect a note of corporate selflessness and concern for the homes of mine workers that would not have been the case in actuality.

As was the case with the Fairmont coal model, the removal of the Pittsburgh model from its exposition context to the United States National Museum deprived its visitors of some visual clues that might have helped them contextualize the model as a product of a deliberate corporate strategy, had they encountered the display in its original setting. In figure 6.7, the Pittsburgh model can be seen in its context at the St. Louis World's Fair of 1904. The case itself is different from the simple museum-standard glass box: the bottom portion of the exposition case jutted outward so as to better highlight the underground part of the model. A large sign on top of the case advertises the company's product as among the best coals "in the world"—a boastful claim recognizable as advertising. A small table with two chairs sits in front of the case, inviting visitors to rest and contemplate the model, or perhaps conduct some important business deal with the company's exposition representative. Two potted plants suggest a kind of informality, and are also a subtle allusion to the origin of coal in fossilized plant matter. Two large columns of the company's coal, bearing square labels, represent viscerally the height of the seams in the mine and invite the visitor to inspect the superior product up close and in person. Thus, in its original exposition context, both the presentation of the Pittsburgh coal model and the objects surrounding it would have forcefully suggested the commercial context and content of the display. Once the model was removed to the museum, the context that rendered the model clearly recognizable as part of a corporate communication strategy disappeared. Instead, though the name of the company remained prominent, the commercial purpose of the model was obscured by a new pedagogical one, backed by the silent authority of the curator of the United States National Museum, whose testimony suggested only the literal truth of the display.

Curator Gilbert began writing to mining and manufacturing companies to find support for his plans shortly after taking the curator job. His letters make absolutely clear that he was willing to provide advertising benefits to companies in exchange for their cooperation and that, ideally, he wanted the companies to develop the exhibits on their own. The museum would merely install them in spaces reserved for individual companies. Gilbert's plan, which seems antithetical to modern conceptions of museum practice, was not inconsistent with the Smithsonian's longstanding involvement in federal sponsorship of scientific and technological investigations with deliberate and positive effects on private industry, as well as other contemporary efforts in the United States National Museum that brought commercial interests into the museum. More to the point, however, Gilbert (and at least some other museum curators and scientists) saw common cause with the heads of companies and industries, in opposition to the public the exhibits proposed to educate, and sought to mobilize the visual culture of mining to help shape public understanding.

Gilbert clearly conceived of his audience as being the sort that might have been present at a world's fair, where, in a mining context, visual displays provided entertainment and education for the public at large, and visitors who were more technically minded could see the latest products displayed by their manufacturers and learn about their technical

Figure 6.7. The same model as figure 6.6, but in its originally intended commercial context at the 1904 St. Louis World's Fair. Objects surrounding the model at the company's booth reflect that the model was part of a corporate marketing effort. From "Coal Mining at Louisiana Purchase Exposition: Description of Models Showing Works of Some of the Large Bituminous Coal Companies," *Mines and Minerals* 25 (September 1904): 81.

capabilities. Gilbert did not have the room to dedicate to full-size machinery, but he was quite interested in achieving a similar effect by creating exhibits to pass on information from companies to visitors, in the forms of models, photographs, and text.

In an attempt to secure the participation of the Johns-Manville Company, manufacturers of asbestos and asbestos products, Gilbert emphasized the advertising value of a display in his museum division, and reminded the company that the museum would bear the brunt of the cost of the display.

> I think you cannot fail to recognize [the proposed exhibit's] decided advantages to you, viewed simply in the light of advertisement. The series would show in full, the industrial uses to which your Company has adapted asbestos, and the system of labeling would set forth your claims as to the merits in each instance, and your name would be conspicuously present. Accordingly, your name and products would be brought conspicuously to the attention of many technical men from all parts of the country daily. No expense would be attached, further than supplying the material, inasmuch as all case work, etc., is provided by the Museum. The only provision which might be regarded as profitless to you would

be that of supplying photographs, etc., calculated to illustrate the technology of your mill and quarry work. Even this feature, however, can scarcely be regarded in that light, since it will serve to bring added interest and attention to your exhibit.[54]

Gilbert's overtures to the mining companies was not simply a byproduct of efforts to secure representative specimens of technology to depict modern practice. Instead, he was quite willing to exhibit even patented machinery made by a single company. For example, he wrote to the Braun Corporation, makers of grinding and crushing equipment, seeking "small models of such products as are covered by [the] basic patent in your name."[55] Gilbert also approached the Deister Machine Company, makers of advanced milling machinery. He wished to discover "whether your Company would care to place in the hall of ore dressing an exhibit setting forth the working principle and advantages of your Tables and Slimer." Gilbert again showed that he was quite willing to display proprietary designs in the museum: he told the company that the exhibit "would be directed simply toward displaying the advantages of your apparatus to the technical public."[56]

Gilbert was willing to cede control of the content of mineral technology exhibits to corporations in an effort to encourage participation. In a letter to the Galigher Machinery Company, which had expressed no interest in contributing to the museum, Gilbert notes that the company would be in control of the exhibit's form and informational content. "The Museum in arranging its exhibits in this hall is offering to make reservations of space in the names of such manufacturers as wish to be represented," but the company did not need to worry about hewing to a strictly defined plan, noted the curator, as "it will be left largely with the exhibitor as to the nature of the exhibit. . . . The purpose of the exhibits will be simply that of enabling such visitors to familiarize themselves with the working principles and the advantages claimed by manufacturers for their various products."[57]

In a handful of circumstances, where Gilbert wanted to ask favors of the largest industrial concerns, the curator asked the secretary of the Smithsonian, Charles D. Walcott, to broach the subject with the industrialists.[58] In 1915, for example, Gilbert wanted to turn to the American Smelting and Refining Company (ASARCO) for exhibit material about lead mining and smelting. Gilbert's contacts at a different company recommended ASARCO as the logical choice, so Gilbert wrote Walcott deferentially, asking if the secretary "would be in a position to broach the subject to Mr. Guggenheim, the President of the Company."[59] Gilbert also used Walcott to approach U.S. Steel.[60]

Perhaps the most important episode where Walcott served as high-powered intermediary was the effort to persuade Daniel C. Jackling, proprietor of Utah Copper Company, creator of the famous Bingham Pit, and father of modern mass production mining, to donate an exhibit to the museum.[61] Such a display as Gilbert wanted would not be inexpensive—a complete model would cost about $3,500.[62] Jackling was clearly interested in the prospect, but after consulting with his other directors, he passed on the opportunity in the short term and asked Walcott to bring it up again later. This the secretary did in early 1916, and this time Jackling was willing to proceed. The copper magnate put Smithsonian officials in touch with R. C. Gemmell, general manager of the Utah property, to work out the details, including a visit by the Smithsonian's choice of model makers. Walcott, in turn, handed the museum side of the arrangements off to Gilbert. The model, at a scale of one inch to forty feet (likely the same forty scale used on the Utah Copper Mine maps), would be sixteen feet by nineteen feet on the floor of the museum exhibit hall. By mid-1917, the model was built and ready to be painted. In preparing the model for exhibit, Gilbert asked for additional material from the company, including a large block of ore, specimens of the country rock, and, echoing Dewey's displays of three decades before, samples of material as the ore passed through each processing stage before becoming copper metal. He also broached the possibility of another addition to the display—a bas-relief panel that would depict the processing of the ore. The model of the pit, large and technologically sublime as it was on its own, did not fully capture Gilbert's pedagogical hopes for the exhibits. The curator noted that he wanted to "impress upon the public the necessity of large scale enterprise in attaining efficiency of development" and, consequently, it was important to fully highlight the "elaborateness" of the operation.[63] In other words, he wanted to convince visitors that massive corporations were not scary unnaturalistic monopolies, but rather were good and necessary institutions for America, because only they could create and master the complex and "efficient" technical operations that generated resources wisely and inexpensively.[64]

Not all companies thought that contributing to the museum would be worthwhile. J. E. Burleson, proprietor of a mica company of the same name in Spruce Pine, North Carolina, replied to Gilbert's circular: "I have no specimens of minerals that I wish to donate to any Museum for advertisements. I have donated several minerals to several Institutions and never got any results from it and consequently I am tired of that kind of business." Burleson did offer to sell Gilbert some uranium and a large amethyst crystal, however.[65]

Gilbert's warm and occasionally sycophantic letters to mineral technology companies suggests his empathy with the industries he exhibited, but his notion of the purpose of public education highlights this property most clearly. Gilbert believed the museum's mission was to educate the public to appreciate the work of industry and to create public acquiescence to monopolistic and efficient use of resources. For him, this sort of education would help prevent unnecessary governmental expansion, would lead to less waste of byproducts, and was an antidote to labor unionism and creeping socialism. Though such ideas are detectable even in Gilbert's early writing, his tone becomes much more strident as the United States attempted to avoid, then became embroiled in, World War I.

In a letter to a university researcher, Gilbert asked for information on the scientist's progress on recovering nitric acid from coal byproducts. (Since nitric acid was a key component of gunpowder, a reliable domestic supply was important for a country nervous about the war raging in Europe.) "The idea is to get together a little exhibit showing the chemical procedure involved in this method of preparing nitric acid, together with the extent of the availabilities that open up and then to issue a descriptive bulletin on the subject." Gilbert described the need for such a plan: "I have been impressed with the thought that the country is running off on an unreasonable tangent in contemplating the erection of a Government controlled hydro electric plant for the fixation of atmospheric nitrogen as being the only practicable means of giving this country its own independent source of nitric acid."[66]

Gilbert's sense of the museum as an inoculation against growing federal control and regulation of industry appears again in a letter to American Cyanamid, offering the company an exhibit at the museum. By the time Gilbert dictated his letter to American Cyanamid, the United States had entered World War I and was embarking on a series of projects to place key industries and industrial sectors under greater federal control. Gilbert described his exhibit idea as "educational propaganda aiming to awaken public opinion" about the need to cultivate a chemical industry that had the freedom to pursue "coordination of opportunity"—in other words, where large corporations did not need to fear antitrust regulators. Gilbert decried the sensational "extension of public authority over industry" stimulated by the crisis of the war. The curator's real concern, however, was not the present state of industry, but what would happen after the war, when a laboring public stimulated by its gains during wartime might demand more. "Things will not revert to the old basis at the close of the war, on the contrary it will be followed by a concentration of public attention upon socialization," worried Gilbert. New exhibits portraying the industry in all its positive complexity were necessary, because "every effort should be made to insure [sic] enlightenment of public opinion as a basis for its demands upon the government for action."[67]

Gilbert occasionally revealed even more clearly his concern that the public would want to unnecessarily regulate all industries because of the actions of a few bad apples. The prospect of such public action made it clear that Gilbert needed to use the educational models and displays at his disposal as a counterweight, to trumpet the importance of industry. As Gilbert argued (with his typical aversion for periods counterbalanced with a undying love for commas), "[W]e have got to allay suspicion, we have got to enlighten society to some true sense of discrimination between merit and demerit in its industrial servants, enabling it to recognize that apart from conspicuously vicious elements industry as a whole is rendering intelligent, industrious, highly efficient service, and that to arbitrarily interfere with this most highly developed of all skilled service is to produce results comparable in a small way to what would follow from arbitrary interference with the cook in the work of preparing meals."[68]

Gilbert's correspondence with the Anaconda Copper Mining Company asking for the company's support for an exhibit and bulletin on the copper industry reveals Gilbert's probusiness thinking. "The mineral industries constitute the economic backbone of the country and the whole economic future of the country is dependent on their efficient development," he wrote, but, he pointed out, efficiency required large investments of capital that made such industries "particularly susceptible to the various economic diseases arising from popular ignorance." As a result, an informed, "appreciative," educated public was "essential to our national growth," and working to create such public opinion was an appropriate task for the museum.[69] In a later letter, Gilbert clarified the need for the sort of "popular education" he had in mind to undertake on "behalf of the mineral industries." Anaconda and other Butte producers were at that time in the middle of a violent fight against radical labor unionism and a paralyzing strike, prompted by a disastrous underground fire in June 1917 that killed 163 miners.[70]

> The purpose of this work is to obviate as far as possible the ill effect of such publicity as that of late accruing to the copper industry. While the direct outcome from such misinterpretations may be inconsiderable, the contribution made to the slow working cumulative influence of public opinion is serious. Whether prompted by misguided sentimentalism or deliberate dishonesty, the result is the same in tending to strengthen the position of the forces of disorganization with a consequent increment to the self confidence and to the general spirit of unrest. The significant eventualities ahead for the labor of mineral industry production are thus bound to represent to an increasing degree the reactions from the shaping of public opinion, and it is therefore important to elevate public comprehension above the plane of sheer instinctive sentiment.[71]

Gilbert never fully realized his vision of a coordinated, integrated set of models and bulletins, covering every domestic mining industry, to spread sanitized and positive information about industrial mining to the public. The mining models at the heart of his exhibits were his most obvious successes, because, as Gilbert noted, their visuality was the "basis of appeal" to average visitors.[72] Some of these models, including the Fairmont and Pittsburgh models described above, had been made by corporations for display at expositions and came to the museum fully formed. Others, such as that of the Utah Copper Company, were funded by companies and subject to their approval, although they were built by museum-affiliated staff. Still others depicted the machinery or facilities of particular corporations, but were created by museum staff following company-provided blueprints at the museum's expense. Gilbert was annoyed by the tendency of museum visitors to look at his models in the wrong order, turning his careful, "systematic" arrangement of facts into a mere "jumble of more or less interesting objects."[73] As a result, he continually tinkered with different forms for his new models, eventually moving away from the free-standing model typified by the Fairmont and Pittsburg models toward linear bas-relief panels that would force patrons to view the sequence of interpretation as Gilbert wanted it.[74] His series of bulletins, ideally based on his exhibits and illustrated with photographs of his models, saw only a handful of numbers issued.[75] Moreover, the disruption caused by World War I to museum operations, including Gilbert's activities, proved to be of permanent consequence. Curators including Gilbert turned their attention to assisting government boards requesting information. Budgets were slashed and salaries regressed as inflation eroded the buying power of frozen wages. Service-age personnel were drafted, and those beyond the reach of the military, from maintenance workers to curators, left the museum in droves because of the miserly compensation, which persisted for years after the war was concluded.[76] Gilbert's staff followed the exodus in 1919. Carl Mitman, an assistant curator (who happened to be Gilbert's brother-in-law), moved to another museum division for a promotion. Joseph Pogue, another mineral technology curator, left the United States National Museum for a position with Sinclair Oil, and in October 1919 Gilbert left for a private sector engineering and consulting career.[77] Mitman was later handed responsibility for the mining collections in addition to his other work, and, predictably, he had little time to devote to them.[78] Gilbert's up-to-date models slowly became obsolete, and in the early 1930s the museum acknowledged the likely need to emphasize the historical nature of the exhibits, because too few resources could be devoted to keeping them current.[79] Ultimately, the visual power inherent in mine models was of tremendous importance in education, whether it was of future mining engineers or the general public.

Before the 1920s, mine models were used in formal educational settings in Europe, but American mining engineering programs suffered from a dearth of models. Companies donated some models, leftovers from lawsuits or expositions, to mining engineering schools. Other models were made by the mining engineering instructors or students. Mining professors recognized the usefulness of visual models in training young engineers; despite their cost, difficulties with their use, and a lack of consensus on the best way to use them in the classroom, these three-dimensional expressions of the visual culture of mining became an important part of the development of new engineers.

Similarly, the world view of mining engineers was brought to public audiences through mine models exhibited at world's fairs and in museums. Prior to the 1890s, mining displays at exhibitions and museums consisted primarily of smaller specimens. Expositions helped drive the creation of permanent exhibits in museums such as the Smithsonian Institution's United States National Museum. Curator Chester Gilbert, in charge of mining displays at the USNM, used mine models extensively to provide, as he put it, "educational propaganda."[80] Many models he used had come directly from corporations, sometimes via expositions, and others were made by the museum according to corporate direction. He used these models to visually convey to the visiting public an understanding of mining that was harmonious with the view of mining engineers. He was also deeply motivated to educate the public to accept, and even be thankful for, the the business strategies and practices of gigantic industrial firms. Gilbert's agenda of boosting the mineral industries, downplaying labor struggles, and fighting off government intervention was visible not only in his letters to mining companies and museum officials, but also, implicitly, in the models that formed the exhibits in the museum. The curator's use of models was deliberate: he chose them for the power of suggestion and coherence that they brought to his displays. In the end, Gilbert recognized—as had mining engineers—the utility of visual tools in a mining context, and used them to support his goal of boosting the American mining industry.

Conclusion

This study ends about 1920. By then the mining engineering profession had undergone tremendous change in the United States, in lockstep with the industrialization of mining. Particularly with the advent of mass mining, engineers had made themselves an essential part of any mining operation—an outcome that might not have been obvious a few decades before.

One element of this transformation was the development and maturation of a visual culture of mining, where maps and models of underground spaces helped mining engineers assert and exercise authority over mining operations. This visual culture had, at its core, two elements. First, it was a set of practices: of creating, using, storing, and revising visual representations. These practices became essential to the work that mining engineers did; learning them became part of their professional training, these practices even became integrated into their professional identity. Second, this visual culture of mining was the material basis of those practices. Every underground map; each beautiful model; and the blueprints, notebooks, transits, and drafting room furniture together represented a piece of the visual culture of mining. Becoming a mining engineer meant learning to make, wield, and understand those tools. They enabled mining engineers to extend their reach beyond the boundaries of the mine plant, to courtrooms, classrooms, and exhibit spaces, to convey the story of industrial mining in the visual language they chose. Though engineers might not always control the reception of their message in venues beyond their immediate control, the visual rhetoric of the representations they shared at least made sure the artifacts projected a message that was implicitly compatible with their purposes. From underground spaces carved out of solid rock to spaces beyond, the visual culture of mining became integrally associated with modern mining operations.

But mining operations do not last forever, as minerals do not regrow themselves in the earth. One of the primary operational premises of industrial mining was to extract the maximum amount of material at the minimum cost, and meanwhile look for other places to mine once the current mine's resources were exhausted. While mining engineers might take their visual knowledge, surveying tools, and drafting equipment with them to new places, what happened to these maps and models as the visual representations grew outdated and mines closed?

The value of saving maps of abandoned mines had been recognized, in fact, before the usefulness of maps in everyday operations was common. Spanish mining regulations governing Mexico in the eighteenth century did not require miners to use maps; if a mine was to be abandoned, however, the authorities would make a map and keep a record of the mine in the official archives.[1] Similar provisions for preserving maps of abandoned mines also sometimes appeared in the early mine safety statutes. Those statutes that required that mines provide inspectors with maps generally also required that a finished map be filed with the same inspectors when the mine was to close (or a portion of the mine was permanently abandoned). Such provisions appeared in both the 1869 and 1870 pioneering Pennsylvania mine safety laws.[2] However, the requirements of such laws, when they existed, were not always fulfilled. An editorial aside in a 1911 *Engineering and Mining Journal* noted that maps ought to be filed with inspectors when a mine closed—implying that, routinely, they were not.[3]

It is easy to understand why. Mining is an optimistic business. A mine might suspend operations for a little bit, just until conditions improved or prices returned to normal. If this temporary shutdown turned permanent, who was left to update the maps and turn them over to the inspectors? Ensuring that maps were finished and submitted meant spending money on engineering talent when the mine had already ceased providing income. As a consequence, the visual representations of a closing mine were commonly neglected. Earlier maps might exist, but as they would seemingly not be useful once the mine was closed, their preservation was often not a priority.

Much as it had once been a site of early and important developments in underground mapping and the professionalization of American mining engineers, in the mid-twentieth century the anthracite coal region of northeastern Pennsylvania prompted important developments in the preservation of the visual culture of mining. By that time, the anthracite industry had been in a state of serious decline for decades, largely replaced as a domestic fuel by more convenient oil or, later, natural gas. A disastrous inundation killed twelve miners and ended mining in an entire district in 1959.[4] Though Pennsylvania elected officials and some residents retained hope that somehow the mines might one day spring back to life, by the mid-1960s it had become increasingly clear that major anthracite mining was a thing of the past.

Unfortunately, the hazardous legacies of anthracite mining were very much part of the present. Old workings filled with water made any attempts at new mining hazardous, and acid-laden water draining from mines caused devastating impacts on riparian environments. Abandoned mines located above the waterline were susceptible to catching fire. Although anthracite mines burned in the historic period as well, it was easier to catch, isolate, and douse a mine fire when the rest of the underground infrastructure was accessible. Without active mining, fighting underground fires became extremely difficult. The best-known example was a fire that apparently started in 1962 in the anthracite town of Centralia, Pennsylvania, that eventually forced evacuation of the entire town. (The Centralia fire continues burning today.)[5] New and existing buildings were threatened by subsidence, where unsupported underground cavities suddenly caved in. The surface movement caused by subsidence could crack foundations, open holes in yards and streets, and swallow cars.

Motivated by concerns about residents' health and safety, along with an ongoing mission to try to assist the ailing anthracite mining industry return to viability, the US Bureau of Mines started a project in June 1963 to collect and microfilm maps of abandoned mines in the anthracite district.[6] This was clearly a rescue operation, an attempt to save potentially valuable engineering data before they were irretrievably lost. Bureau personnel intended to begin microfilming maps and records from the Eastern Middle anthracite field first, but "it appeared prudent to microfilm first the maps of all mines that appeared on the verge of being abandoned," resulting in a smattering of maps from across northeastern Pennsylvania receiving priority treatment.[7] In initial publications describing their efforts, Bureau personnel highlighted the precarious circumstances of the records of many abandoned mines. When a mine closed, maps might end up in the hands of estate attorneys, retired engineers, or the widows of mining engineers. Potentially worse, the records might remain at the mine, succumbing to the elements along with the rest of the mine infrastructure. "At another mine office the maps were destroyed by fire. Fortunately, the maps of this particular mine had been microfilmed by the Bureau under this program about a month prior to the fire."[8]

The choice of microfilm as the storage medium seemed clear to the Bureau. While having the actual drawings would be the best overall solution, acquiring, storing, and caring for these visual representations would be a great burden. By contrast, once microfilmed, all the maps could be stored "in a single filing cabinet," and easily duplicated.[9] The Bureau used a large copy stand arrangement with both backlights and top-mounted lights that could be adjusted to meet the needs of the different map materials. Two standard reduction ratios were used, to permit reproduction to scale if needed. Each mine was numbered, and every map from that mine was numerically indexed as well, with the number recorded in the microfilm camera shot. Many maps were so large that even at a thirty-to-one reduction ratio they had to be captured piece by piece over several frames.[10]

To secure cooperation of mine operators, the Bureau took several steps. It targeted abandoned mines particularly. Staff filmed maps of active mines, but promised not to include them in the published lists of maps until the mines were closed. Most important, staff assured mine owners that the recorded maps were only for "technical" or other scientifically oriented studies. The microfilms were explicitly not for "any taxation, investigation, regulation, or litigation purposes," and this promise was secured by signed confidentiality agreements.[11]

The Bureau began filming in June 1963, and within a year had captured the maps of eighty-seven mining operations—forty-three in the Eastern Middle anthracite field and the rest across northeastern Pennsylvania.[12] By the beginning of 1967, the Bureau had microfilmed the maps from 210 mines; it issued a second catalog of maps, covering much of the Northern field, in 1968.[13] Two more reports followed, bringing the total to 8,348 maps from 317 major and independent anthracite mines by August 1970.[14] Further work progressed more slowly. The report on the Southern field, issued in 1978, detailed that 8,615 maps from 323 major and independent mines had been microfilmed, which the report's author estimated covered about 97 percent of all anthracite mines.[15] A planned summary report, which would have included once-active mines and any missed by the intervening reports, was apparently never completed.

Though the anthracite map filming project lasted into the late 1970s, by the end of the 1960s the potential promise it showed of a cooperative program to microfilm mine maps was enough to spur the Bureau of Mines to create two formal mine map repositories and embark on extensive efforts to gather and catalog historic underground maps. The two repositories in Denver and Pittsburgh opened in 1970, covering mines in the Western and Eastern parts of the country, respectively.[16] Like the earlier anthracite efforts, these new repositories would borrow and microfilm mine maps, then return them to their owners. In contrast to the efforts needed to scrounge up anthracite maps, the new mine map repositories took advantage of their wider scope to pursue readily available maps first. State-level geological surveys and mining bureaus were the source of the majority of maps microfilmed in the early years, though some companies also made privately held maps available for filming.[17] New federal regulations to enhance coal mine safety, passed in 1969, required maps to be made and furnished to federal authorities as well, spurring an influx of coal mining maps to the Pittsburgh repository.[18]

The new repositories also made use of microfilm as the archival medium of choice. After a series of maps were filmed on to a reel of microfilm, a copy was made and chopped into single frames. These individual microfilm chips were then mounted into punchcard-style aperture cards that also contained the document number of the map. Information about the map, such as location and date, was entered into a computerized database. A set of aperture cards was provided back to the map donor when the maps were returned after microfilming, and a set was kept at the repository.[19]

At first, the need to save the maps at risk of being lost meant that fewer efforts were placed on how to use maps that had been saved. Initial reproductions were photographic enlargements from the microfilm.

These scaled effectively because of the care initially taken to record scale when the maps were first photographed. Using a process akin to enlarging a photo, staff could reproduce a map on paper or a transparent layer that they could then add to other maps. In the late 1970s repository staff worked out procedures to use computers to laboriously digitize mine boundaries from individual maps, and even electronically plot the maps on the character printers of the day, using different letters to represent different areas of coverage on the printed map.[20]

As part of massive budget cuts in 1983, the Bureau of Mines lost funding for the repositories. The Office of Surface Mining Reclamation and Enforcement (OSMRE), part of the Department of the Interior, adopted the Pittsburgh and Wilkes-Barre repositories that October, and were able to continue the work of collecting and preserving mine maps (concentrating largely on the coal mines prevalent in the East). The Denver and Spokane repositories were mothballed and did not microfilm any new maps. The repositories had microfilmed more than 90,000 maps before the budget problems halted the work in the West.[21] When the Bureau of Mines was permanently abolished in 1996, the contents of the Denver and Spokane repositories were transferred to the Pittsburgh-area repository, which correspondingly was given a nationwide scope of operations.[22]

Less than a decade later, the Quecreek miners narrowly escaped with their lives from an underwater flood caused by breaking through to an abandoned mine. The fact that an updated map had been found in a nearby mining museum indicated how valuable checking abandoned mine maps could be for the safety of new operations—and also showed that valuable information clearly still existed outside the repository's control, placing miners and mines at risk.[23] In response, the Mine Safety and Health Administration (MSHA) announced nearly $4 million in grants in 2004 to help states find and digitize historic mine maps, resulting in a flood of new information to the National Mine Map Repository (NMMR) in Pittsburgh. The repository handled the larger scanning jobs directly on its state-of-the-art equipment. The MSHA grants ended in 2011.[24] The repository now records historic maps in digital format as well as archival microfilm, and can provide copies of maps on DVDs instead of photo prints, but the essence of recording the visual culture of mining engineers, whether from the nineteenth or twentieth century, continues much as it had in decades prior.

In the twenty-first century, efforts have been made to save historic mine maps—the products of the visual culture of mining that developed during the industrialization of mining in the late nineteenth and early twentieth centuries—because the legacies of the mines they represent are still with us. Though the mining engineers who made them have long since passed, the spaces they created are very much still part of our lives. Once the mines close, their maps are what can help us visualize historic mining spaces, perhaps to imagine new exploitation of old deposits to meet future energy and mineral needs, or possibly to help avoid subsidence, inundation, and other hazards. But whatever our present or future purposes, these maps and models were tools and not static pictures. They allow us to see underground mining landscapes through the eyes of the engineers who made both tools and mines. Without their efforts, the vast workings below the surface of the earth would forever remain invisible to us.

NOTES

Introduction

1. Quecreek Miners, *Our Story: 77 Hours that Tested Our Friendship and Our Faith,* as told to Jeff Goodell (New York: Hyperion, 2002).

2. "Investigating Grand Jury Report No. 1," Supreme Court of Pennsylvania 208 M.D. Misc. Dkt. 2001; Court of Common Pleas, Dauphin County, Pennsylvania, No. 660 M.D. 2001, p. 6; copy in possession of author. Many thanks to Peter Liebhold for obtaining this document for me.

3. "Investigating Grand Jury Report No. 1," 6–12.

4. "Investigating Grand Jury Report No. 1" ["lying in a corner" (7), "inexplicably" (8)].

5. See, especially, James W. Cook, "Seeing the Visual in U.S. History," *Journal of American History* 95, no. 2 (2008): 432–441; and Margaret Dikovitskaya, *Visual Culture: The Study of the Visual after the Cultural Turn* (Cambridge, MA: MIT Press, 2005).

6. Elspeth H. Brown, *The Corporate Eye: Photography and the Rationalization of American Commercial Culture, 1884–1929* (Baltimore: Johns Hopkins University Press, 2005), 15.

7. T. A. Rickard, *A History of American Mining* (New York: McGraw-Hill, 1932), 2–11 ["by the time George Washington" (11)].

8. Rickard, *A History of American Mining,* 18–29.

9. See Rickard, *A History of American Mining;* Duane A. Smith, *Mining America: The Industry and the Environment, 1800–1980* (Lawrence: University Press of Kansas, 1987, reprint, Niwot: University Press of Colorado, 1993); and Rodman W. Paul and Elliott West, *Mining Frontiers of the Far West, 1848–1880,* 2nd ed. (Albuquerque: University of New Mexico Press, 2001).

10. Rosalind Williams, *Notes on the Underground: An Essay on Technology, Society, and the Imagination* (Cambridge, MA: MIT Press, 1990), 4–8.

11. Gregg S. Clemmer, *American Miners' Carbide Lamps: A Collector's Guide to American Carbide Mine Lighting* (Tucson, AZ: American Miners' Carbide Lamps: A Collector's Guide to American Carbide Mine Lighting, 1987); Mark Aldrich, *Safety First: Technology, Labor, and Business in the Building of American Work Safety, 1870–1939* (Baltimore: Johns Hopkins University Press, 1997), 219, 224–229.

12. To say that the literature on technological systems is deep is undoubtedly an

understatement, but the work of Thomas P. Hughes is generally regarded as the start-
ing point, especially *Networks of Power: Electrification in Western Society, 1880–1930*
(Baltimore: Johns Hopkins University Press, 1983); and Thomas P. Hughes, "The
Evolution of Large Technological Systems," in *The Social Construction of Technological
Systems: New Directions in the Sociology and History of Technology*, ed. Wiebe E. Bijker,
Thomas P. Hughes, and Trevor Pinch (Cambridge, MA: MIT Press, 1987), 51–82.
Hughes regarded mines as "system artifacts" which were natural things adapted to
fit large-scale systems (Hughes, "The Evolution," 51). By contrast, historian Thomas
Andrews terms underground mines *workscapes*, emphasizing that they are a product
of labor interacting with nature and vice versa. Thomas G. Andrews, *Killing for Coal:
America's Deadliest Labor War* (Cambridge, MA: Harvard University Press, 2008), esp.
123–127. Timothy J. LeCain analyzes underground mines as part of an "envirotech-
nical" system, a move intended to blur distinctions between natural and artificial
environments. *Mass Destruction: The Men and Giant Mines That Wired America and
Scarred the Planet* (New Brunswick, NJ: Rutgers University Press, 2009), esp. 10–11,
22, 37–38.

13. The foundations of the study of mining engineers by historians were laid by
Clark Spence in his 1970 work, Clark C. Spence, *Mining Engineers and the American
West: The Lace-Boot Brigade, 1849–1933* (reprint, Moscow: University of Idaho Press,
1993), esp. 2–4 ["focal figure" (4)]; also see Rodman Wilson Paul, "Colorado as a
Pioneer of Science in the Mining West," *The Mississippi Valley Historical Review* 47,
no. 1 (June 1960): 34–50.

14. Logan Hovis and Jeremy Mouat, "Miners, Engineers, and the Transformation of Work in the Western Mining Industry, 1880–1930," *Technology and Culture*
37, no. 3 (July 1996): 429–456, describe and analyze this shift ["progressive and
widespread" (430), "mass production" (433)]. Historian Timothy LeCain innova-
tively twists the term, arguing we should recognize nonselective mining as "mass
destruction."

15. Andrew Abbott, *The System of Professions: An Essay on the Division of Expert
Labor* (Chicago: University of Chicago Press, 1988). This would appear to be a little
different from the protracted battles fought between the managers and owners of
mines, on one hand, and mine labor on the other, which could and did escalate to
levels of remarkable brutality on both sides. Engineers were always on management's
team, but as often as not they stood on the sidelines during actual conflict. This element of mining history has received much well-deserved attention. Andrews, *Killing for Coal*, is an award-winning and well-nuanced account of the coalfield wars
of Colorado. Academic studies more sympathetic to labor include Mark Wyman,
Hard Rock Epic: Western Miners and the Industrial Revolution, 1860–1910 (Berkeley:
University of California Press, 1979); Richard E. Lingenfelter, *The Hardrock Miners:
A History of the Mining Labor Movement in the American West, 1863–1893* (Berkeley:
University of California Press, 1974); Vernon H. Jensen, *Heritage of Conflict: Labor*

Relations in the Nonferrous Metals Industry up to 1930 (Ithaca, NY: Cornell University Press, 1950); George Suggs, *Colorado's War on Militant Unionism* (Norman: University of Oklahoma Press, 1972); and Donald L. Miller and Richard E. Sharpless, *The Kingdom of Coal: Work, Enterprise, and Ethnic Communities in the Mine Fields* (Philadelphia: University of Pennsylvania Press, 1985). Historian Brian Frehner frames an analysis of the development of the oil industry as a professionalization struggle as well, between educated geologists and practical oil men, played out in dialog and tension with the oil-bearing landscape. Brian Frehner, *Finding Oil: The Nature of Petroleum Geology* (Lincoln: University of Nebraska Press, 2011).

16. Eugene S. Ferguson, "The Mind's Eye: Nonverbal Thought in Technology," *Science* 197 (August 1977): 827–836; he further developed the implications of his argument in Eugene S. Ferguson, *Engineering and the Mind's Eye* (Cambridge, MA: MIT Press, 1992). Also see Walter G. Vincenti, *What Engineers Know and How They Know It: Analytical Studies from Aeronautical History* (Baltimore: Johns Hopkins University Press, 1990); Brooke Hindle, *Emulation and Invention* (New York: New York University Press, 1981); and Kathryn Henderson, *On Line and On Paper: Visual Representations, Visual Culture, and Computer Graphics in Design Engineering* (Cambridge, MA: MIT Press, 1999).

17. Steven Lubar, "Representation and Power," *Technology and Culture* 36, no. 2, Special Issue (1995): S54–S81; esp. S62–S63 ["technological representations" (S54–S55), "enormous power" (S62)]; also John K. Brown, "When Machines Became Gray and Drawings Black and White: William Sellers and the Rationalization of Mechanical Engineering," *IA: The Journal of the Society for Industrial Archeology* 25, no. 2 (1999): 29–54.

18. "[T]he engineering drawing allows the engineer to master the machine . . . you can dominate a piece of paper in a way that you cannot dominate a machine." Lubar, "Representation and Power," S70.

1 ✕ Underground Mine Maps

1. Denis Wood, *The Power of Maps* (New York: Guilford Press, 1992); J. Brian Harley, *The New Nature of Maps: Essays in the History of Cartography*, ed. Paul Laxton (Baltimore: Johns Hopkins University Press, 2001).

2. Mott T. Greene, *Geology in the Nineteenth Century: Changing Views of a Changing World* (Ithaca, NY: Cornell University Press, 1982), esp. 122–125.

3. Simon Winchester, *The Map That Changed The World: William Smith and the Birth of Modern Geology* (New York: HarperCollins, 2001).

4. J. P. Lesley, *Manual of Coal and its Topography* (Philadelphia: J. B. Lippincott, 1856), 191.

5. That is, the sum of the measurements of the interior angles of a polygon (in degrees) equals 180 times the number of sides of the polygon minus two.

6. Much of the following discussion of mine surveying instruments is drawn

from the exhaustively technical and impressively detailed accounts on the topic initially published in the *Transactions* of the AIME in the last years of the nineteenth century and collected together as Dunbar D. Scott and others, *The Evolution of Mine-Surveying Instruments* (New York: American Institute of Mining Engineers, 1902).

7. A modern "theodolite" is essentially a transit, but the older name has persisted.

8. Philip A. Arnold, "Surveying and Mapping," *Coal Age* 12 (1917): 1109–1110.

9. W. L. Owens, "Surveying and Mapping," *Coal Age* 12 (1917): 960–961.

10. Arnold, "Surveying and Mapping," 1110.

11. Eckley B. Coxe, "Remarks on the Use of the Plummet-Lamp in Underground Surveying," *Transactions of the American Institute of Mining Engineers* 1 (1873): 379.

12. See, e.g., Scott and others, *The Evolution of Mine-Surveying Instruments,* 62.

13. Coxe, "Remarks on the Use of the Plummet-Lamp in Underground Surveying."

14. Andro Linklater, *Measuring America: How the United States Was Shaped by the Greatest Land Sale in History* (New York: Plume Books / Penguin, 2002).

15. Eckley B. Coxe, "Improved Method of Measuring in Mine Surveys," *Transactions of the American Institute of Mining Engineers* 2 (1875): 219–224.

16. Robert Peele, ed., *Mining Engineers' Handbook,* 1st ed. (New York: John Wiley and Sons, 1918), 1249–1250.

17. On the companies, see Larry D. Lankton, *Hollowed Ground: Copper Mining and Community Building on Lake Superior, 1840s–1990s* (Detroit, MI: Wayne State University Press, 2010); Larry D. Lankton, *Cradle to Grave: Life, Work, and Death at the Lake Superior Copper Mines* (New York: Oxford University Press, 1991); Larry D. Lankton and Charles K. Hyde, *Old Reliable: An Illustrated History of the Quincy Mining Company* (Hancock, MI: Quincy Mine Hoist Association, 1982). A useful older corporate history is C. Harry Benedict, *Red Metal: The Calumet and Hecla Story* (Ann Arbor: University of Michigan Press, 1952). On the office buildings themselves, see Renee M. Blackburn, "Preserving and Interpreting the Mining Company Office: Landscape, Space, and Technological Change in the Management of the Copper Industry" (M.S. Thesis, Michigan Technological University, 2011). The Michigan Technological University Archives and Copper Country Historical Collections hold a large amount of excellent primary source material for both C&H and Quincy; the Keweenaw National Historical Park Archives are smaller and less comprehensive, but they care for extensive collections of original artifacts.

18. See, for example, the large Calumet and Hecla (1871) map and the later Coxe Drifton maps described in chapter 2. "Workings of Calumet and Hecla Mine, 1871," Box B1, Calumet and Hecla Unprocessed Drawings, Michigan Technological University Archives, Houghton, MI; "Drifton No. 2. Buck Mountain Vein." Map 14-10, and "Drifton Slope No. 2, Gamma Vein." Map 14-11, both in Collection 1002 Coxe Brothers Collection, Archives Center, National Museum of American History, Smithsonian Institution (hereafter AC).

19. I draw this from examples in the Quincy Mining Company Collection, MS-001, Copper Country Historical Collections and Michigan Technological University Archives, Houghton, MI.

20. For examples, see drawings 5710, 6959, and 7667, drawer 99, folders A–B, MS-005 Calumet and Hecla Drawing Collection, Michigan Technological University Archives.

21. This piece survives in the collection of the Keweenaw National Historical Park, Calumet, MI, which cares for surviving C&H structures.

22. Arnold, "Surveying and Mapping," 1110.

23. A fathom, six feet, was the standard unit of measurement in Cornwall, so it is not a surprise to find fathom-derived measurements, particularly in mining districts where the Cornish had a strong influence on mining practice.

24. The effect of blueprints in concentrating power centrally in the hands of engineers would seem to function differently than the office duplication technologies such as carbon copies examined by historian JoAnne Yates, who saw the latter as generally having a "centrifugal," decentralizing force in a firm. JoAnne Yates, *Control through Communication: The Rise of System in American Management* (Baltimore: Johns Hopkins University Press, 1989), 62.

25. P. Barnes, "Note upon the So-Called Blue Process of Copying Tracings, Etc.," *Engineering News* (October 1878); P. Barnes, "Note upon the 'Blue' Process of Copying Tracings Etc.," *Transactions of the American Institute of Mining Engineers* 6 (1879): 197–198; Ernst Lietze, *Modern Heliographic Processes: A Manual of Instruction in the Art of Reproducing Drawings, Engravings, Manuscripts, Etc., by the Action of Light; for the Use of Engineers, Architects, Draughtsmen, Artists, and Scientists* (New York: D. Van Nostrand, 1888); Benjamin James Hall, *Blue Printing and Modern Plan Copying for the Engineer and Architect, the Draughtsman and the Print Room Operative* (London: Sir I. Pitman & Sons, Ltd., 1921).

26. George E. Brown, *Ferric and Heliographic Processes: A Handbook for Photographers, Draughtsmen, and Sun Printers* (New York: Tennant & Ward, 1900), 59–60; "Cross Hatching for Stope Maps, Quincy Mining Co." and "Cross Hatching— Ground Plan Maps," both in Box 08, Collection MS-012 "Quincy Mining Company Engineering Drawings Collection," Michigan Technological University Archives.

27. The earliest mention of a law with this proviso that I have found was Wyoming's 1886 law, which used the term *blue print*; see Sir Frederick Augustus Abel, *Mining Accidents and Their Prevention: With Discussion by Leading Experts* (New York: Scientific, 1889), 342. By 1895 and 1903 Pennsylvania and Tennessee also had similar language in their mine safety laws, though they used the term *sun print*.

28. James T. Beard, *Mine Examination Questions and Answers,* 1st ed. (New York: McGraw-Hill, 1923), 485.

29. Owens, "Surveying and Mapping," 960.

30. Eckley B. Coxe, "Technical Education," *Transactions of the American Society of Mechanical Engineers* 15 (1894): 661.

31. H. G. Reist, "Blue-Printing by Electric Light," *Transactions of the American Society of Mechanical Engineers* 22 (1901): 897.

32. Yates, *Control through Communication,* 28–31.

33. H. L. Botsford, "An Index System for Maps," *Engineering and Mining Journal* 94 (1912): 638.

34. Rufus J. Foster, "Protect Mine Maps and Plans," *Coal Age* 10 (1916): 376.

35. Botsford, "An Index System for Maps."

36. "Classification and Method of Card Indexing Maps and Drawings, Coxe Bros. & Co., (1929)" folder "Coxe Collection Mining Map Drwgs," drawer "Coxe Collection (Administrative)," Division of Work and Industry, National Museum of American History, Smithsonian Institution (hereafter DWI).

37. "Pigeon Holes," Box 08, Collection MS-012 "Quincy Mining Company Engineering Drawings Collection," Michigan Technological University Archives.

38. Georgius Agricola, *De Re Metallica,* trans. Herbert Clark Hoover and Lou Henry Hoover (London: The Mining Magazine, 1912), 3–4.

39. Agricola, *De Re Metallica,* 128–148.

40. Agricola, *De Re Metallica,* 129; Hoover's comment appears at the bottom of n. 16.

41. Scott and others, *The Evolution of Mine-Surveying Instruments,* 9–10.

42. Scott and others, *The Evolution of Mine-Surveying Instruments,* 147.

43. John S. Hittel, *Hittel on Gold Mines and Mining* (Quebec [Quebec City?]: G. & G. E. Desbarats, 1864), 147.

44. J. Ross Browne, *Letter from the Secretary of the Treasury Transmitting A Report upon the Mineral Resources of the States and Territories West of the Rocky Mountains,* 39th Cong, 2d Sess., House of Representatives Ex. Doc. No. 29 (Washington, DC: GPO, 1867), 235.

45. "Map of Mamoth [sic] Vein Workings in Greenfield & Temperance Basins, Beaver Meadow Collieries, Accessibel [sic] For Surveys, Of Old Workings of 1830 to 1864 No Maps Exist," Map 2-22, Collection 1002 Coxe Brothers Collection, AC.

46. Hittel, *Hittel on Gold Mines and Mining,* 38.

47. Scott and others, *The Evolution of Mine-Surveying Instruments,* 146.

48. Scott and others, *The Evolution of Mine-Surveying Instruments,* 132, 221–223, 242.

49. Myron Angel, ed., *History of Nevada, with Illustrations and Biographical Sketches of Its Prominent Men and Pioneers* (Oakland, CA: Thompson and West, 1881), 587; James B. McNair, *With Rod and Transit: The Engineering Career of Thomas S. McNair (1824–1901)* (Los Angeles: Published by the Author, 1951).

50. H. L. Fickett, "Mine Maps vs. Mine Sketches," *Coal Age* 8 (1915): 741.

51. Bituminous or soft coal mines in Pennsylvania were regulated under a

different set of laws, similar in terms and scope to the anthracite laws but with a handful of important differences, passed several years after the original anthracite legislation. Frank F. Brightly, ed., *Brightly's Purdon's Digest: A Digest of the Statute Law of the State of Pennsylvania from the Year 1700 to 1894*, 12th ed., vol. 2 (Philadelphia: Kay and Brother, 1894), 1345–1346, 1359–1360; Albert H. Fay, *Coal-Mine Fatalities in the United States, 1870–1914*, US Bureau of Mines Bulletin 115 (Washington: GPO, 1916), 284–286, 296.

52. "Importance of Accurate Mine Surveys," *The Colliery Engineer* 8 (May 1888), 228.

53. "Importance of Accurate Mine Surveys."

54. Fickett, "Mine Maps vs. Mine Sketches," 741.

55. *Third Biennial Report of the State Mine Inspectors, to the Governor of the State of Iowa, for the Years 1886 and 1887* (Des Moines, IA: Geo. E. Roberts, State Printer, 1888), 78.

56. B. W. Robinson, "Mine Maps: A Paper Showing What Is Required on a Good Map, and Why Complete Maps Are a Necessity," *The Colliery Engineer* 9 (April 1889): 198. Robinson was engineer for the St. Bernard Coal Company.

57. Robinson, "Mine Maps" [all quotes on 198].

58. Robinson, "Mine Maps" [all quotes on 198].

59. Fickett, "Mine Maps vs. Mine Sketches" [all quotes on 741].

60. Fickett, "Mine Maps vs. Mine Sketches," 740.

61. F. B. Richards, "Mining," in *Engineering as a Career: A Series of Papers by Eminent Engineers,* ed. F. H. Newell and C. E. Drayer (New York: D. Van Nostrand, 1916), 153.

62. Owens, "Surveying and Mapping," 960.

63. Fickett, "Mine Maps vs. Mine Sketches" [all quotes on 740].

64. James G. McGivern, "Polytechnic College of Pennsylvania: A Forgotten College," *Journal of Engineering Education* 52, no. 2 (November 1961): 106.

65. McGivern, "Polytechnic College of Pennsylvania," 108; Thomas Thornton Read, *The Development of Mineral Industry Education in the United States* (New York: American Institute of Mining and Metallurgical Engineers, 1941), 35–39.

66. McGivern, "Polytechnic College of Pennsylvania," 106, 111.

67. Read, *Development of Mineral Industry Education.*

68. McGivern, "Polytechnic College of Pennsylvania," 110.

69. Spence, *Mining Engineers and the American West*, 30–32, 37–38, 50.

70. Spence, *Mining Engineers and the American West*, 70–78.

71. Eckley B. Coxe, "Presidential Address, Proceedings of the Lake George and Lake Champlain Meeting, October 1878," *Transactions of the American Institute of Mining Engineers* 7 (1879): 103–114 [all quotes on 107].

72. H. S. Munroe, "A Summer School of Practical Mining," *Transactions of the American Institute of Mining Engineers* 9 (1881): 664–671.

73. T. A. Rickard, *Interviews with Mining Engineers* (San Francisco: Mining and Scientific Press, 1922), 255.

74. R. H. Richards, "American Mining Schools," *Transactions of the American Institute of Mining Engineers* 15 (1887): 309–340; R. H. Richards, "American Mining Schools," *Transactions of the American Institute of Mining Engineers* 15 (1887): 809–819. The latter is an updated supplement to Richards' first article.

75. Bennett Brough, *A Treatise on Mine-Surveying,* London: Charles Griffin; Philadelphia: J. B. Lippincott, 1888, 3, reported these were Nicolaus Voigtel, *Geometria Subterranea* (1686); followed by the works of J. F. Weidler (in Latin, 1726); H. Beyer, (in German, 1749); and von Oppel (in German, 1749) Brough reported that the earliest authors to address the subject in English were Thomas Houghton in 1686, William Pryce in 1778, and Thomas Fenwick in 1804, but all of these works were relatively rare by the middle of the nineteenth century.

76. George G. Hood, "Mine Surveying" (C. E. thesis, Lehigh University, 1883), Lehigh University Library Special Collections, Thesis 1883 H777m.

77. Brough, *A Treatise on Mine Surveying, 1888.* I have found an edition published in 1926, thirty-eight years after the first, and eighteen years after Brough's death in 1908. Bennett H. Brough and Harry Dean, *A Treatise on Mine-Surveying* 17th ed., rev by Henry Louis (London: C. Griffin and Company, 1926).

78. According to the "Preface to the First Edition" dated 1886, J. B. Johnson's treatise on surveying included a chapter on mine surveying written by a US Deputy Mineral Surveyor. The chapter was completely rewritten by different authors, one a professor at the Colorado School of Mines, for the fifteenth edition, issued in 1900 (J. B. Johnson, *The Theory and Practice of Surveying,* 16th ed. [New York: John Wiley & Sons, 1908], xii, iv–v).

79. Reuben Street, "Mine Surveying," *The Colliery Engineer* 8 (March 1888): 177.

80. Fickett, "Mine Maps vs. Mine Sketches," 741. Throughout Fickett's article, he notes the responsibility of the mining engineer to the company that employs him, and he saw good maps as a function of good and lasting service to an employer by the engineering profession. Historian Edwin T. Layton has noted that by the turn of the twentieth century, the mining engineers were most closely associated in their professional identity with the businesses for which they worked, compared to the other major branches of American engineering. See Edwin T. Layton, *The Revolt of the Engineers: Social Responsibility and the American Engineering Profession* (Cleveland, OH: The Press of Case Western Reserve University, 1971).

81. George L. Yaste, "Care of Mine Maps," *Coal Age* 10 (1916) [all quotes on 223].

82. Marionne Cronin noted that the airplane, which was the tool of survey geologists and mining engineers searching for new deposits in Canada, became an icon for that group as well. See Marionne Cronin, "Northern Visions: Aerial Surveying and the Canadian Mining Industry, 1919–1928," *Technology and Culture* 48 (April 2007): 303–330.

2 ✕ Anthracite Mapping and Eckley Coxe

1. Rossiter W. Raymond, "Biographical Notice of Eckley B. Coxe," *Transactions of the American Institute of Mining Engineers* 25 (1896): 451–454.

2. Independent producers were those not directly controlled by a railroad company.

3. Colliery refers to the total mining plant, including the underground mine and surface processing facilities.

4. On anthracite mining, see Miller and Sharpless, *The Kingdom of Coal*; John N. Hoffman, *Anthracite in the Lehigh Region of Pennsylvania, 1820–45*, Contributions from the Museum of History and Technology 72 (Washington, DC: Smithsonian Institution Press, 1968); and Anthony F. C. Wallace, *St. Clair: A Nineteenth-Century Coal Town's Experience with a Disaster-Prone Industry* (Ithaca, NY: Cornell University Press, 1988).

5. *Reports of the Inspectors of Coal Mines of the Anthracite Coal Regions of Pennsylvania for the Year 1878* (Harrisburg, PA: Lane S. Hart, State Printer, 1879), 221–224 (hereafter *Anthracite Inspectors' Reports*). Coxe later mined the other coal seams on his property.

6. J. Price Wetherill, "An Outline of Anthracite Coal Mining in Schuylkill County, Pa.," *Transactions of the American Institute of Mining Engineers* 5 (1877): 402–422, which includes the discussion that followed the paper. H. M. Chance, *Report on the Mining Methods and Appliances Used in the Anthracite Coal Fields* (Harrisburg, PA: Second Geological Survey of Pennsylvania, 1883) is the most comprehensive account of 1880s practice. For a thorough but more modern description, see Hudson Coal Company, *The Story of Anthracite* (New York: Author, 1932), 111–181.

7. Richard P. Rothwell, "The Mechanical Preparation of Anthracite," *Transactions of the American Institute of Mining Engineers* 3 (1875): 134–144; Eckley B. Coxe, "The Iron Breaker at Drifton, With a Description of Some of the Machinery Used for Handling and Preparing Coal at the Cross Creek Collieries," *Transactions of the American Institute of Mining Engineers* 19 (1891): 398–474; Coxe described his iron breaker design in exhaustive detail, with forty-two plates, for a technical audience. Also see Chance, *Mining Methods and Appliances*, 457–474; and Hudson Coal Company, *Story of Anthracite,* 182–209, for a description of a 1920s–1930s–era breaker.

8. "Wilkes-Barre Meeting, May 16, 1871," *Transactions of the American Institute of Mining Engineers* 1 (1873): 3–4.

9. "An Act for the Better Regulation and Ventilation of Mines, and for the Protection of the Lives of the Miners in the County of Schuylkill," in *Laws of the General Assembly of the State of Pennsylvania Passed at the Session of 1869* (Harrisburg, PA: B. Singerly, State Printer, 1869), 852–856; also "An Act Providing for the Health and Safety of Persons Employed in Coal Mines," in *Laws of the General Assembly of the State of Pennsylvania Passed at the Session of 1870* (Harrisburg, PA: B. Singerly, State Printer, 1870), 3–12; Wallace, *St. Clair,* 293–305; Aldrich, *Safety First,* 67–68; Robert

P. Wolensky and Joseph M. Keating, *Tragedy at Avondale: The Causes, Consequences, and Legacy of the Pennsylvania Anthracite Coal Industry's Most Deadly Mining Disaster, September 6, 1869* (Easton, PA: Canal History / Technology Press, 2008).

10. Eric Nystrom, "'Without Doubt the Most Accurate': Underground Surveying and the Development of Mining Engineering in the Pennsylvania Anthracite Region," *Pennsylvania Legacies* 9, no. 2 (November 2009): 20–25.

11. Brough, *A Treatise on Mine-Surveying, 1888*, v.

12. Charles A. Ashburner, "New Method of Mapping the Anthracite Coal-Fields of Pennsylvania," *Transactions of the American Institute of Mining Engineers* 9 (1881): 507.

13. "Map of Mamoth [*sic*] Vein Workings in Greenfield & Temperance Basins, Beaver Meadow Collieries, Accessibel [*sic*] For Surveys, Of Old Workings of 1830 to 1864 No Maps Exist," Map 2-22, Collection 1002 Coxe Brothers Collection, AC.

14. "An Act for the Better Regulation and Ventilation of Mines."

15. "An Act Providing for the Health and Safety of Persons Employed in Coal Mines," 3–4.

16. *Reports of the Inspectors of Coal Mines of the Anthracite Coal Regions of Pennsylvania for the Year 1870.* (Harrisburg, PA: B. Singerly, State Printer, 1871), 9, 93.

17. The inspector for the Southern District of Luzerne County, John T. Evans, did not mention maps or their absence in his report for 1870. See *Anthracite Inspectors' Reports for 1870*, 239–250.

18. *Anthracite Inspectors' Reports for 1870*, 230–231. Note that there are more "total surface openings" in his district (seventy-two total surface openings) but since Williams notes that there were "11 shafts working" that had not complied, I use his figures for shafts specifically (230).

19. *Anthracite Inspectors' Reports for 1870*, 235.

20. *Anthracite Inspectors' Reports for 1870*, using figure on p. 230 of 72 "total surface openings" and mention on p. 235 of "about a dozen" unmapped mines.

21. *Reports of the Inspectors of Coal Mines of the Anthracite Coal Regions of Pennsylvania for the Year 1871* (Harrisburg, PA: B. Singerly, State Printer, 1872), 72.

22. *Anthracite Inspectors' Reports for 1871*, 50. There was a large industry-wide strike that shut down operations in the anthracite country for half of 1871, but as regular workers did not conduct the surveys, this should not have made much of an impact on mapping practices. Anecdotal evidence suggests it was not uncommon to catch up on engineering work, especially underground, at off-times such as on Sunday, during holidays, and during strikes, which might have nudged the compliance rate upward.

23. *Anthracite Inspectors' Reports s for 1871*, 89.

24. *Reports of the Inspectors of Mines of the Anthracite Coal Regions of Pennsylvania for the Year 1873* (Harrisburg, PA: Benjamin Singerly, State Printer, 1874), 10, lists the names of 159 Schuylkill District collieries, but the list of collieries in the district

provided on pp. 34–39 lists 166 entries. The list of collieries furnishing maps on pp. 73–74 includes a total of 126, but the paragraph immediately below the table notes that 125 maps were then on file and safely stored. Thus possible rates of compliance vary from 75.30 percent (125 of 166) to 79.25 percent (126 of 159). Note that the following year's report gives a higher figure for number of collieries in 1873 (176), which would skew the compliance rate downward to approximately 71–72 percent.

25. *Anthracite Inspectors' Reports for 1873*, 34–39.

26. *Reports of the Inspectors of Mines of the Anthracite Coal Regions of Pennsylvania for the Year 1874* (Harrisburg, PA: B. F. Meyers, State Printer, 1875), 65, 82, 102, 134, 135, 165. For data about map provision also given for the Eastern District of the Wyoming coal field in Luzerne County, see *Anthracite Inspectors' Reports for 1874*, 166.

27. *Reports of the Inspectors of Mines of the Anthracite Coal Regions of Pennsylvania for the Year 1875* (Harrisburg, PA: B. F. Meyers, State Printer, 1876), 29–34, 42–44.

28. "A Supplement to an Act, entitled 'An Act providing for the health and safety of persons employed in coal mines,'" in *Laws of the General Assembly of the State of Pennsylvania Passed at the Session of 1876* (Harrisburg, PA: B. F. Meyers, State Printer, 1876), 130.

29. *Reports of the Inspectors of Mines of the Anthracite Coal Regions of Pennsylvania for the Year 1876* (Harrisburg, PA: B. F. Meyers, State Printer, 1877.), 49.

30. *Reports of the Inspectors of Mines of the Anthracite Coal Regions of Pennsylvania for the Year 1877* (Harrisburg, PA: Lane S. Hart, State Printer, 1878), 132.

31. *Anthracite Inspectors' Reports for 1877*, 132.

32. *Anthracite Inspectors' Reports for 1877*, 131.

33. *Anthracite Inspectors' Reports for 1877*, 131.

34. Raymond, "Biographical Notice of Eckley B. Coxe"; Philip B. McDonald, "Eckley Brinton Coxe," *Dictionary of American Biography* Base Set, American Council of Learned Societies, 1928–1936, Reproduced in *Biography Resource Center* (Farmington Hills, MI: Thomson Gale, 2007); James J. Bohning, "Angel of the Anthracite: The Philanthropic Legacy of Sophia Georgina Coxe," *Canal History and Technology Proceedings* 24 (2005): 150–182.

35. For the first decades of its operation, the colliery itself was called the Cross Creek Colliery, abbreviated C.C.C., and was run by Coxe Brothers. Coxe Brothers was a partnership until after Eckley Coxe's death, when it incorporated, then was sold to the Lehigh Valley Coal Company. However, the estate of Tench Coxe did not sell its lands, so even after the corporation was sold Tench Coxe's heirs continued to receive royalties on coal mining into the 1930s. See Patrick Shea, "Against All Odds: Coxe Brothers & Company, Inc. and the Struggle to Remain Independent" (Paper delivered at the Mining History Association Conference, Scranton, PA, June 2005).

36. "A Pleasant Excursion," *New York Times*, October 19, 1890, 2.

37. Hunter Rouse, "Weisbach, Julius Ludwig," in *Dictionary of Scientific Biography*,

ed. Charles Coulston Gillispie, vol. 14 (New York: Charles Scribner's Sons, 1976), 232.

38. Julius Weisbach, *Theoretical Mechanics: With an Introduction to the Calculus: Designed as a Text-Book for Technical Schools and Colleges, and for the Use of Engineers, Architects, Etc.; Translated from the Fourth Augmented and Improved German Edition,* trans. Eckley B. Coxe (New York: D. Van Nostrand, 1875).

39. Julius Weisbach, *Neue Markscheidekunst und ihre Anwedung auf Bergmännische Anlagen* (Braunschweig: Verlag von Friedrich Vieweg und Sohn, 1859).

40. A. B. Parsons, "History of the Institute," in *Seventy-five Years of Progress in the Mineral Industry, 1871–1946,* ed. A. B. Parsons (New York: American Institute of Mining and Metallurgical Engineers, 1947), 496.

41. *Proceedings of the American Philosophical Society* 11 (1870): 521; *Proceedings of the American Philosophical Society* 12 (1871): 1; "Condensed Dispatches," *Washington Post,* January 12, 1894, 7.

42. Eckley B. Coxe, Heber S. Thompson, and William Griffith, *Report of the Commission Appointed to Investigate the Waste of Coal Mining, With the View to the Utilizing of the Waste* (Philadelphia: Allen, Lane & Scott's Printing House, May 1893); Eckley B. Coxe, "A Furnace With Automatic Stoker, Travelling Grate, and Variable Blast, Intended Especially for Burning Small Anthracite Coals," *Transactions of the American Institute of Mining Engineers* 22 (1894): 581–606; Eckley B. Coxe, "Some Thoughts upon the Economical Production of Steam, With Special Reference to the Use of Cheap Fuel," *Transactions of the New England Cotton Manufacturers' Association* 58 (1895): 133–210.

43. James J. Bohning, "Chemistry, Coal and Culture: The Library of Eckley Brinton Coxe" (Paper delivered at the National Meeting of the American Chemical Society, Chicago, August 2001); H. C. Bradsby, ed., *History of Luzerne County, Pennsylvania, with Biographical Selections* (Chicago: S. B. Nelson & Co., 1893), 304.

44. "A Conscience in Pennsylvania Politics," *Harper's Weekly* February 12, 1881, 99; Bradsby, *History of Luzerne County, Pennsylvania,* 304–307.

45. See notebooks 1–5, Eckley Brinton Coxe Manuscripts (SC MS 059), Lehigh University Library Special Collections.

46. Coxe, "Remarks on the Use of the Plummet-Lamp in Underground Surveying," 378–379. Coxe's design was a significant improvement on a lamp designed by Julius Weisbach, Coxe's professor at Freiberg. Scott and others, *The Evolution of Mine-Surveying Instruments,* 283–284, 303–304. A plummet lamp of this sort is described (without attribution) in the standard mining engineer's handbook of the twentieth century; see Robert Peele and John A. Church, eds., *Mining Engineers' Handbook,* 3rd ed., vol. 2 (New York: John Wiley & Sons, Inc., 1941), 18-04. [Note: Peele's page references are section-page format.]

47. Coxe, "Improved Method of Measuring in Mine Surveys"; Scott and others,

The Evolution of Mine-Surveying Instruments, 32, 284–285. Also see Peele and Church, *Mining Engineers' Handbook,* 18-14, 18-15.

48. *Anthracite Inspectors' Report for 1878,* 224.

49. Bohning, "Angel of the Anthracite"; Eckley B. Coxe, "Secondary Technical Education," *Transactions of the American Institute of Mining Engineers* 7 (1879): 217–226; *Anthracite Inspectors' Report for 1878,* 221–224; "[A] better feeling" (224).

50. House Select Committee on Existing Labor Troubles in Pennsylvania, *Labor Troubles in the Anthracite Regions of Pennsylvania, 1887–1888,* 50th Congress, 2nd Session, House of Representatives Report 4147 (Washington, DC: GPO, 1889), 592.

51. "There May Be Trouble," *Washington Post,* November 20, 1887, 1; "Quiet in the Coal Regions," *Washington Post,* November 21, 1887, 1.

52. House Select Committee on Existing Labor Troubles in Pennsylvania, *Labor Troubles,* 599.

53. House Select Committee on Existing Labor Troubles in Pennsylvania, *Labor Troubles,* 598, 602 ["I have never turned" (598)].

54. *Coxe Brothers & Company* v. *The Lehigh Valley Railroad Company* 4 I.C.C. 535 (1891).

55. "No Coal War Expected," *New York Times,* February 11, 1889, 8.

56. Bradsby, *History of Luzerne County, Pennsylvania,* 297.

57. "Present Workings of the Cross Creek Colliery, Drifton, Luzerne County, Pennsylvania, October 1st. 1870, Scale 100 feet to the inch," Map 14-22, Collection 1002 Coxe Brothers Collection, AC.

58. Eckley B. Coxe, *Mining Legislation,* author's offprint, Lehigh University Special Collections (Philadelphia: A Paper Read at the General Meeting of the American Social Science Association, October 1870), 5.

59. Coxe, *Mining Legislation,* esp. 16–18.

60. "Tracing of the C.C.C. Workings Slope No. 1 showing extension of working [*sic*] April 1872," Map 14-23, Collection 1002 Coxe Brothers Collection, AC.

61. "Map of a Portion of the Cross Creek Colliery, Luzerne County Pa, 1879," Map 14-27, Collection 1002 Coxe Brothers Collection, AC.

62. The average elevation for the town of Drifton, Pennsylvania, is 1,660 feet above sea level, according to the USGSm's Geographical Names Information Service (accessed online).

63. *Anthracite Inspectors' Reports for 1870,* 80–83; *Anthracite Inspectors' Reports for 1871,* 81–83.

64. At least twelve individual maps bound together at the top form Map 14-16, Collection 1002 Coxe Brothers Collection, AC. This analysis concentrates on the map marked "Sheet 2 of 12."

65. On Eckley Coxe's death, see fn34 this chapter; "All Quiet at Hazleton," *New York Times,* September 15, 1897, 3; "Windup of the Coal Strike," *New York Times,*

September 21, 1897, 3; "Martin on the Stand," *Washington Post*, March 2, 1898, 1; "Bottom of Strike," *Washington Post*, June 10, 1902, 1; "Mitchell as Peacemaker," *Washington Post*, July 4, 1902, 8; "More Coal Mined," *Washington Post*, October 25, 1902, 1; "Talk of Mine Strike," *Washington Post*, November 11, 1902, 1; "Lehigh Road Announces Coxe Brothers Sale," *New York Times* October 14, 1905, 13.

66. "Drifton No. 2. Buck Mountain Vein, West End Workings, Bottom Bench, Scale 1" = 100 ft., Revised to 3-1-1919, Revised to 11-24-19, Revised to 5-1-1921, Revised to 1-24-1923, revised to March 1925, Revised to Dec. 1925, Revised to Dec 1927," Map 14-10, Collection 1002 Coxe Brothers Collection, AC.

67. The gangway is for haulage, one of the main arteries of the mine, and generally close to level. The manway is a small passageway up into the working breast, probably with a ladder, that the miners walk through to get to their working place. It is similar to the street in front of your house (gangway) and the stairs you have to climb to your front door (manway).

68. E. Kudlich to L. C. Smith, August 2, 1897, folder 13, box 703, collection 3005, Coxe Family Mining Papers, Historical Society of Pennsylvania, Philadelphia (hereafter HSP)

69. Authorization #4340, folder 6, "Drifton, 1923–1936," Box 737, "Pillar Mining Authorizations, Drifton Colliery, 1923–1936," Collection 3005, Coxe Family Mining Papers, HSP.

70. For example, E. Kudlich, Mining Engineer, Drifton, to Wm. Jones, Mine-Boss, Tomhicken, February 28, 1900, folder 13, box 703, coll. 3005, HSP.

71. Lubar, "Representation and Power"; Ferguson, "The Mind's Eye."

72. On the history of mining engineering and its spread worldwide, see Spence, *Mining Engineers and the American West*.

3 ✕ New Maps, the Butte System, and Geologists Ascendant

1. Isaac F. Marcosson, *Anaconda* (New York: Dodd, Mead, 1957), 19–66. Marcosson's work is a celebratory corporate history of Anaconda, but remains the only comprehensive history of the firm and can be informative if the reader makes sufficient allowance for its origins.. An excellent history of Butte that largely ends with the organization of the modern Anaconda Company is Michael P. Malone, *The Battle for Butte: Mining and Politics on the Northern Frontier, 1864–1906* (Seattle: University of Washington Press, 1981). Elements of Butte's labor, ethnic, and gender history have been explored fruitfully by historians, including David Emmons and Mary Murphy; see David M. Emmons, *The Butte Irish: Class and Ethnicity in an American Mining Town, 1875–1925* (Urbana: University of Illinois Press, 1989); and Mary Murphy, *Mining Cultures: Men, Women, and Leisure in Butte, 1914–41* (Urbana: University of Illinois Press, 1997).

2. Gordon Morris Bakken, *The Mining Law of 1872: Past, Politics, and Prospects* (Albuquerque: University of New Mexico Press, 2008).

3. Charles Meyer et al., "Ore Deposits at Butte, Montana," in *Ore Deposits of the*

United States, 1933–1967: The Graton-Sales Volume, vol. 2, ed. John D. Ridge (New York: American Institute of Mining, Metallurgical, and Petroleum Engineers, Inc., 1968), 1373–1416.

4. Walter Harvey Weed, *Geology and Ore Deposits of the Butte District, Montana,* USGS Professional Paper 74 (Washington: GPO, 1912), 15.

5. A contributing factor was the different scales on which geological theories and mining operations acted. Knowing that several hundred square miles had been covered by lava eons ago did little to help the miner guess if there might be gold hidden in the rocks forty-five feet ahead.

6. Hittel, *Hittel on Gold Mines and Mining,* 28.

7. A broad, contemporary overview is provided by T. A. Rickard, "Geology Applied to Mining," *Mining and Scientific Press* 100 (1910): 479–481.

8. Rodman W. Paul and Elliott West, *Mining Frontiers of the Far West, 1848–1880,* 2nd ed. (Albuquerque: University of New Mexico Press, 2001), 130–132; Samuel Franklin Emmons, *Geology and Mining Industry of Leadville, Colorado,* USGS Monograph 12 (Washington, DC: GPO, 1886).

9. "David William Brunton," *Mining and Metallurgy* 9 (January 1928): 37; "Brunton Awarded First Mining Medal," *Mining and Metallurgy* 8 (February 1927): 82–83; Ginny Kilander, "Transits, Timbers, & Tunnels: The Legacy of Colorado Inventor David W. Brunton," in *Enterprise and Innovation in the Pikes Peak Region,* ed. Tim Blevins (Colorado Springs, CO: Pikes Peak Library District with Dream City Vision 2020, 2011), 65–101.

10. D. W. Brunton, "A New System of Ore-Sampling," *Transactions of the American Institute of Mining Engineers* 13 (1885): 639–645.

11. David W. Brunton, "Engineering Record" [typescript], vol. 1, pp. 13–14, Western History Collection, Denver Public Library, Colorado.

12. "Engineering Record," vol. 1, pp. 25–40. Brunton told T. A. Rickard he worked for Daly until the mine owner sold out to Amalgamated, which was correct but misleading. Rickard, *Interviews with Mining Engineers,* 79; "Brunton Awarded First Mining Medal," 82, uses similar language. In fact, his diary-like "Engineering Record," which at times provides a day-by-day picture of his activities, indicates Brunton spent considerable time consulting for Amalgamated after Daly's death and resigned as consulting engineer only in 1905.

13. Malone, *The Battle for Butte,* 45–46.

14. Rickard, *Interviews with Mining Engineers,* 79.

15. Reno H. Sales, "Ore Discovery and Development of Fundamental Importance: How Anaconda's Large Geological Staff Is Employed," *Engineering and Mining Journal* 128 (1929): 277.

16. Rickard, *Interviews with Mining Engineers,* 501–504.

17. "Proceedings of the Lake Superior Meeting," *Transactions of the American Institute of Mining Engineers* 27 (1898): xxx, xxxv–xxxvii.

18. Rickard, *Interviews with Mining Engineers,* 505.

19. Horace V. Winchell, "The Work of the Geological Department of the Amalgamated Copper Company" [typescript and manuscript], 1906, folder "Reports—Horace V. Winchell, 1898–1913," Newton Horace Winchell and Family Collection, P453, Box 10, Minnesota Historical Society, 3.

20. F. A. Linforth, "Application of Geology to Mining in the Ore Deposits at Butte Montana," in *Ore Deposits of the Western States: The Lindgren Volume,* ed. John Wellington Finch and The Committee on the Lindgren Volume (New York: The American Institute of Mining and Metallurgical Engineers, 1933), 695.

21. Malone, *The Battle for Butte* [quote on 189]. More problematic, but useful because of interviews with contemporaries of the struggle, is C. B. Glasscock, *War of the Copper Kings: Builders of Butte and Wolves of Wall Street* (New York: Bobbs-Merrill, 1935).

22. F. Ernest Richter, "The Amalgamated Copper Company: A Closed Chapter in Corporation Finance," *Quarterly Journal of Economics* 30, number 2 (February 1916): 387–407; also Malone, *The Battle for Butte,* 205–206. I am grateful to Fred Quivik for bringing the Richter article to my attention.

23. Malone, *The Battle for Butte* ["unscrupulous ingenuity" (189)].

24. Reno H. Sales, *Underground Warfare at Butte* (Caldwell, ID: Caxton Printers, 1964); also Malone, *The Battle for Butte,* 179–181.

25. Rickard, *Interviews with Mining Engineers,* 79–80.

26. David W. Brunton, "Geological Mine-Maps and Sections," *Transactions of the American Institute of Mining Engineers* 36 (1906): 508.

27. Brunton, "Geological Mine-Maps and Sections," 508–509, 539–540.

28. Brunton, "Geological Mine-Maps and Sections," 509.

29. Winchell, "The Work of the Geological Department," 5.

30. Brunton, "Geological Mine-Maps and Sections," 510.

31. These included a qualitative description of the ore, a note on how wide the sample was when taken from the mine, and assay returns. Brunton, "Geological Mine-Maps and Sections," 510.

32. Brunton, "Geological Mine-Maps and Sections," 510, 540. These procedures gradually evolved within the company, but the broad methods and purposes remained the same. See F. A. Linforth and E. B. Milburn, "Geology Applied to Mining," *Engineering and Mining Journal* 91 (April 1911): 664–667; J. L. Hayes, "Anaconda Copper Mining Co.'s Mapping Practice," *Engineering and Mining Journal-Press* 116 (1923): 841–843; and Linforth, "Application of Geology to Mining in the Ore Deposits at Butte Montana."

33. Brunton, "Geological Mine-Maps and Sections" [all quotes on 539]. Linforth, "Application of Geology to Mining in the Ore Deposits at Butte Montana," "engineering accuracy" (696).

34. Winchell, "The Work of the Geological Department," 5.

35. Winchell, "The Work of the Geological Department," 5.

36. Malone, *The Battle for Butte,* 187.

37. Rickard, *Interviews with Mining Engineers,* 509.

38. Vincent D. Perry and Charles Meyer, "Reno H. Sales," in *Ore Deposits of the United States, 1933–1967: The Graton-Sales Volume,* vol. 1, ed. John D. Ridge (New York: The American Institute of Mining, Metallurgical, and Petroleum Engineers, Inc., 1968), xxi–xxiii.

39. Winchell, "The Work of the Geological Department," 7.

40. "Interview with Reno Sales by Henry Carlisle, Butte, Montana, January 1961" [typescript], Columbia University Oral History Research Office, Mining Engineering Project, No. 437, Volume 3, 15.

41. Sales Oral History, 16.

42. The article does not say what company Gow works for, but he worked for Anaconda from his graduation in 1907 from the Colorado School of Mines until at least 1911. "Personals: '07," *The Colorado School of Mines Magazine* 1, no. 9 (June, 1911): 16. In May 1911, he became city engineer for Butte. Within a few years Gow was superintendent of the Pilot Butte Mining Co., one of the few non-Anaconda mines left in Butte by the 1910s, which promptly became embroiled in an apex lawsuit with Anaconda. See "Pilot Butte Mining Co., Montana," *The Copper Handbook* 11 (1914): 715–716.

43. Localized mine coordinate systems were used in Butte from the arrival of Brunton and Winchell, but were eventually consolidated into a single district-wide coordinate grid no later than 1923. See Hayes, "Anaconda Copper Mining Co.'s Mapping Practice."

44. Paul A. Gow, "Mine Surveying Methods Employed at Butte, Montana," *Engineering and Mining Journal* 90 (1910): 1209.

45. Gow, "Mine Surveying Methods," 1209–1210.

46. Given that the Brunton pocket transit had a clinometer built-in, Gow may have used this tool for the rough vertical measurements he described. Gow, "Mine Surveying Methods," 1210–1211.

47. Gow, "Mine Surveying Methods," 1210–1211.

48. Gow, "Mine Surveying Methods," 1211.

49. Gow, "Mine Surveying Methods," 1211.

50. Peter H. von Bitter, "The Brunton Pocket Transit, a One Hundred Year Old North American Invention," *Earth Sciences History* 14, no. 1 (1995): 98–102; David W. Brunton, "Pocket-Transit," US Patent 526,021, filed March 10, 1894, and issued September 18, 1894.

51. Folder 2, Collection +83, Newton Horace Winchell and Family Papers, Minnesota Historical Society, St. Paul, MN (hereafter Winchell Family Papers).

52. Linforth and Milburn, "Geology Applied to Mining," 664–665.52. Winchell Family Papers, "Foot wall" (Map 9); "Can This" (Map 500); "This can be considered" (Map 7); "CuSO" (Map 500); "Should follow" (Map 500); "Enargite" (Map 500).

53. Later, Anaconda geologists would use specialized notebooks to record information underground, which presumably would have been saved for long-term use. Linforth and Milburn, "Geology Applied to Mining," 664–665.

54. Frank A. Linforth, "Applied Geology in the Butte Mines," *Transactions of the American Institute of Mining Engineers* 46 (1914): 110–122.

55. Linforth, "Applied Geology in the Butte Mines"; the example is repeated, with additional details but a slightly different numbering system, in Linforth, "Application of Geology to Mining in the Ore Deposits at Butte Montana," 696.

56. Linforth, "Applied Geology in the Butte Mines," 122.

57. J. H. Farrel, "Geological Mapping of Mine Workings," *Engineering and Mining Journal* 86 (1908): 385.

58. Sales, "Ore Discovery and Development," 277.

59. Rickard, *Interviews with Mining Engineers,* 80–81.

60. Winchell, "The Work of the Geological Department," 7 [emphasis in original].

61. Rickard, *Interviews with Mining Engineers,* 80–81.

62. Sales Oral History typescript, 13–14 [both quotes on 14].

63. David W. Brunton, "Modern Progress in Mining and Metallurgy in the Western United States," *Transactions of the American Institute of Mining Engineers* 40 (1910); [all quotes on 544].

64. Brunton mentioned that the economic geologists were "not burdened with the duties of surveying, directing workmen, etc." which by implication were job responsibilities of mining engineers that absorbed time that might otherwise be spent on big-picture planning (Brunton, "Modern Progress," 544). This resonates with the work of historian Kathleen Ochs, who notes the evolution of the mining engineer's work into a management role. See Kathleen H. Ochs, "The Rise of American Mining Engineers: A Case Study of the Colorado School of Mines," *Technology and Culture* 33, no. 2 (April 1992): 278–301.

65. "Brunton Awarded First Mining Medal"; [all quotes on 82].

66. L. C. Graton, "Seventy-five Years of Progress in Mining Geology," in *Seventy-five Years of Progress in the Mineral Industry, 1871–1946,* ed. A. B. Parsons (New York: American Institute of Mining and Metallurgical Engineers, 1947), 14–15.

67. Sales, "Ore Discovery and Development," 279.

4 ✕ Modeling the Underground in Three Dimensions

1. Scientific models have been attracting considerable attention from historians of science in recent years. For example, see Mary S. Morgan and Margaret Morrison, eds., *Models as Mediators: Perspectives on Natural and Social Science* (Cambridge, UK: Cambridge University Press, 1999), which emphasizes theoretical models over physical artifacts; and Soraya de Chadarevian and Nick Hopwood, eds., *Models: The Third*

Dimension of Science (Stanford, CA: Stanford University Press, 2004), which covers both types but tends to emphasize models as material artifacts.

2. Ludmilla Jordanova, "Material Models as Visual Culture," in *Models: The Third Dimension of Science,* ed. Soraya de Chadarevian and Nick Hopwood (Stanford, CA: Stanford University Press, 2004), 443.

3. Eric Francoeur, "The Forgotten Tool: The Design and Use of Molecular Models," *Social Studies of Science* 27 (1997) ["embody, rather than" (14), "models mimic" (16)].

4. Molecular models were by no means the only physical models used in a scientific context. Other common examples were models of excavations made by archaeologists, wax anatomical models and moulages, and models of the hulls of ships. Many of these are analyzed by the essays in Chadarevian and Hopwood, *Models;* see also Christine Keiner, "Modeling Neptune's Garden: The Chesapeake Bay Hydraulic Model, 1965–1984," in *The Machine in Neptune's Garden: Historical Studies on Technology and the Marine Environment,* ed. Helen M. Rozwadowski and David K. van Keuren (Sagamore Beach, MA: Science History Publications, 2004), 273–314.

5. Ferguson, *Engineering and the Mind's Eye.*

6. Ferguson, *Engineering and the Mind's Eye,* 102–107.

7. Indeed, the model bears a distinct resemblance to the five-compartment shaft created by the Tamarack Mining Company, which was later bought out by C&H. Item 16, Acc. KEWE-00040 "Grid Rack (Wood)," Keweenaw National Historical Park, Calumet MI; compare with H. M. Lane, "Plumbing Deep Shafts of the Tamarack Mine," *Mines and Minerals* 22 (January 1902): 248.

8. B. H. Dunshee, "Shaft Sinking in Butte, Montana," *Mines and Minerals* 27 (1907): 262–263.

9. Philip B. Bucky, "Use of Models for the Study of Mining Problems," *American Institute of Mining and Metallurgical Engineers Technical Publication,* no. 425 (1931): 3–28; W. H. Craig, "Geotechnical Centrifuges: Past, Present, and Future," in *Geotechnical Centrifuge Technology,* ed. R. N. Taylor (Glasgow: Blackie Academic / Chapman and Hall, 1995), 1–3.

10. The photograph is found in the John Hoffman Curatorial Files, DWI. The mine was reopened as a tourist attraction in the 1980s.

11. C. L. Severy, "Making Mine Models," *Mining and Scientific Press* 104 (1912): 381.

12. John R. Chamberlin, in discussion following H. H. Stoek, "Mine Models," *Transactions of the American Institute of Mining Engineers* 58 (1918): 32.

13. Harrison and Zulch, 50.

14. T. S. Harrison and H. C. Zulch, "Court Maps and Models," *Mines and Minerals* 29 (September 1908): 49–50.

15. Sopwith had a long and prosperous career in the British mining industry

and became quite well recognized before his death in 1879. See Thomas J. Bewick, "Memoir of Mr. Thomas Sopwith," *Transactions of the North of England Institute of Mining and Mechanical Engineers* 29 (1880): 105–111; Sopwith's models may be seen on the website of the Whipple Museum of the History of Science, of Cambridge University. http://www.hps.cam.ac.uk/whipple/explore/models/geologicalmodels/.

16. T. Sopwith, *A Treatise on Isometrical Drawing, as Applicable to Geological and Mining Plans, Picturesque Delineations of Ornamental Grounds, Perspective Views and Working Plans of Buildings and Machinery, and to General Purposes of Civil Engineering; With Details of Improved Methods of Preserving Plans and Records of Subterranean Operations in Mining Districts* (London: John Weald, Taylor's Architectural Library, 1834), 154–155.

17. Sopwith, *A Treatise on Isometrical Drawing,* 154–157.

18. Brough, *A Treatise on Mine-Surveying, 1888,* 263.

19. Thomas Sopwith, *Description of a Series of Geological Models, Illustrating the Nature of Stratification, Valleys of Denudation, Succession of Coal Seams in the Newcastle Coal Field, the Effects Produced by Faults or Dislocation of the Strata, Intersection of Mineral Veins, &c.* (Newcastle-upon-Tyne: Printed for the Author by J. Blackwell, 1841).

20. T. Sopwith, *Description of a Series of Elementary Geological Models* (London: R. J. Mitchell & Sons, 1875), ix–x, 1–2.

21. The King survey, an important predecessor of the USGS, used contours. See Thurman Wilkins, *Clarence King: A Biography,* Rev. Ed. (Albuquerque: University of New Mexico Press, 1988), 110.

22. Benjamin Smith Lyman, "On the Importance of Surveying in Geology," *Transactions of the American Institute of Mining Engineers* 1 (1873): 190–192.

23. O. B. Harden, "Topographical and Geological Modelling," *Transactions of the American Institute of Mining Engineers* 10 (1882): 264–267.

24. E. A. Byler and Lee W. Davis, "Topographic Model of Cripple Creek District," *Mining and Scientific Press* 107 (1913): 144. Note that Byler and Davis are writing some thirty years after the initial work of O. B. Harden was published; they make no mention of the earlier pioneers, but the same techniques and problems of vertical exaggeration are still operative.

25. John H. Harden and Edward B. Harden, "The Construction of Maps in Relief," *Transactions of the American Institute of Mining Engineers* 16 (1888): 294.

26. Harden and Harden, "The Construction of Maps in Relief," 294–295.

27. J. P. Lesley, "Topographical Models or Relief-Maps," *Science* 7, no. 154 (1886): 58.

28. Harden and Harden, "The Construction of Maps in Relief," 298–299.

29. Lyman, "On the Importance of Surveying," 187–188, 192.

30. Lesley, "Topographical Models or Relief-Maps"; Harden and Harden, "The Construction of Maps in Relief," 290.

31. Harden and Harden, "The Construction of Maps in Relief," 298, notes one

underground model, and possibly another, among the fourteen they built for Lesley's pedagogical purposes with the assistance of his students in 1873.

32. Ashburner died young after a promising initial career. See J. P. Lesley, "Biographical Notice of Charles A. Ashburner," *Transactions of the American Institute of Mining Engineers* 18 (1890): 365–370.

33. Charles A. Ashburner, *First Report of Progress in the Anthracite Coal Region: The Geology of the Panther Creek Basin or Eastern End of the Southern Field,* Second Geological Survey of Pennsylvania, Report AA (Harrisburg, PA: Board of Commissioners for the Second Geological Survey, 1883): 242.

34. A. E. Lehman, "Topographical Models: Their Construction and Uses," *Transactions of the American Institute of Mining Engineers* 14 (1886): 453; Frederic P. Dewey, *A Preliminary Descriptive Catalogue of the Systematic Collections in Economic Geology and Metallurgy in the United States National Museum* (Washington, DC: GPO, 1891), 224.

35. Ashburner, quoted in Lehman, "Topographical Models," 454. However, Ashburner certainly knew about Lesley's 1865 underground models mentioned above, which suggests that Ashburner's model was more extensive and useful for solving mining problems than Lesley's 1865 attempt had been.

36. Brough, *A Treatise on Mine-Surveying, 1888,* 262.

37. The model has a custom box lid with a carrying handle that fits over the model to protect it, and is marked "Return to Pottsville Office, P. & R. C. & I. Co." Collections of the DWI.

38. This model is on display at the World Museum of Mining, Butte, Montana.

39. D. W. Brunton, "Aspen Mountain: Its Ores and their Mode of Occurrence," *Engineering and Mining Journal* 46 (1888): 22–23, 42–45.

40. W. I. Evans, "A New Method of Making Mine Models," *Engineering and Mining Journal* 58 (1894): 293.

41. *Annual Report of the Board of Regents of the Smithsonian Institution . . . for the Year Ending June 30, 1889: Report of the National Museum* (Washington: GPO, 1891), 114, 414, 783.

42. Harrison and Zulch, "Court Maps and Models," 51.

43. Mack C. Lake, "Mine Models Made of Celluloid Sheets," *Engineering and Mining Journal* 99 (April 1915): 737–738, 957; "Collapsible Geological Model," *Engineering and Mining Journal* 99, no. 12 (1915): 532–533.

44. Harrison and Zulch, "Court Maps and Models," 50.

45. "Mine Model of Vertical Glass Plates," *Engineering and Mining Journal* 99 (January 1915): 236–237.

46. F. M. Kurie, "Glass Models of Portland Mine: A Convenient Means for Plainly Illustrating Mine-Workings, Geological Features, Etc.," *Mines and Minerals* 24 (February 1904): 307.

47. Control later passed to British capitalists, then through a series of lawsuits

to an American company. George F. Becker, *Geology of the Quicksilver Deposits of the Pacific Slope,* USGS Monograph 13 (Washington, DC: GPO, 1888), 8–10.

48. David J. St. Clair, "New Almaden and California Quicksilver in the Pacific Rim Economy," *California History* 73 (1994): 281–284.

49. Becker, *Geology of the Quicksilver Deposits,* 10–11. New Almaden rebounded after this period and had some very good subsequent years.

50. For a detailed and reliable account of the time between the discovery of the Big Bonanza in 1873 and its termination in 1877, see Grant H. Smith and Joseph V. Tingley, *The History of the Comstock Lode* (Reno: Nevada Bureau of Mines and Geology, in association with University of Nevada Press, 1998), 145–215.

51. Jimmie Schneider, *Quicksilver: The Complete History of Santa Clara County's New Almaden Mine* (San Jose, CA: Zella Schneider, 1992), 67.

52. Lynn R. Bailey, *Supplying the Mining World: the Mining Equipment Manufacturers of San Francisco, 1850–1900* (Tucson, AZ: Westernlore Press, 1996), 21–22. Exhibitions were held at yearly intervals beginning in 1857.

53. "An Interesting Mine Model," *Mining and Scientific Press* 29 (August 1874): 136.

54. "An Interesting Mine Model" [all on 136].

55. Becker, *Geology of the Quicksilver Deposits,* 321.

56. Rickard, *Interviews with Mining Engineers,* 227.

57. Correspondence between Randol and Dewey: J. B. Randol letter to the director of the National Museum, May 14, 1888; Memorandum from [G.B. Goode?] to F. P. Dewey with Dewey's reply, n.d.; J. B. Randol letter to G. Brown Goode, June 2, 1888; RU 305 [microfilm], Smithsonian Institution. Many thanks to Shari Stout for securing copies of these documents for me. Catalogue citation: Dewey, *Preliminary Descriptive Catalogue,* 191, 193. The model was listed as item 66456.

58. "The Eureka-Richmond Case," *Engineering and Mining Journal* 24 (September 1877): 181–182 [quote on 181].

59. *Transcript of Record, U.S. Supreme Court, Nos. 636 (116) and 637 (117), The Richmond Mining Company of Nevada, et al., Appellants, v. The Eureka Consolidated Mining Company, Appeals from the Circuit Court of the United States of the District of Nevada,* filed January 17, 1878, Records and Briefs of the Supreme Court, 355–357, 362–369.

60. "Mine Model," *Mining and Scientific Press* 34 (June 1877): 337.

61. *Eureka-Richmond* transcript, 98, 132.

62. Mary Hallock Foote, *A Victorian Gentlewoman in the Far West: The Reminiscences of Mary Hallock Foote,* ed. Rodman W. Paul (San Marino, CA: Huntington Library, 1972), 164.

63. Schneider, *Quicksilver,* 67–68.

64. C. M. Campbell, "Methods of Mining the Granby Orebodies," *Engineering and Mining Journal* 87 (January 1909): 255.

65. Given Hill's status and wealth as a smelting magnate and US senator, it seems

likely that Chamberlin was the inventor of the technique, and Hill bankrolled its development. The US patent was No. 727,140; "Mine Model or Exhibit," *Engineering and Mining Journal* 75 (1903): 757. (Note: Chamberlin's name is misspelled in that article.) The British patent was number 1,410 of 1903; "Mine Model," *Engineering and Mining Journal* 75 (1903): 904.

66. Patent Specification No. 727,140, 3–4.

67. Chamberlin comment during discussion, in Stoek, "Mine Models," 32.

68. Bennett H. Brough, *A Treatise on Mine-Surveying,* 12th ed. (London: Charles Griffin., 1906), 304.

69. Brough, *A Treatise on Mine-Surveying, 1906,* 303.

70. John W. Meier, "Mines on the Lahn in Nassau, Germany," *Engineering and Mining Journal* 54 (1892): 414–415.

71. The pre-1900 comprehensive maps of the Quincy Mine and the C&H Mine in Michigan's Copper Country offer fine examples, though the practice seems to have been widespread in the period.

72. Edmund D. North, "Glass Mine-Models," *Transactions of the American Institute of Mining Engineers* 40 (1910): 755–759.

73. Maxwell C. Milton, "A Mine Model," *Mining and Scientific Press* 105 (1912): 338.

74. Volney Averill, "Details and Cost of a Mine Model," *Engineering and Mining Journal* 108 (1919): 824–825.

75. "Geologic Mine Model," file 2224, Anaconda Geological Collection, American Heritage Center, University of Wyoming.

76. See, for example, "Wood and Cloth Mine Model," *Engineering and Mining Journal* 95 (February 1913): 419–420.

77. See, for example, the model presumed to be from the *Rarus-Johnstown* case, World Museum of Mining, Butte, Montana.

78. "Solid wood mine model of Anaconda-Neversweat Mine," World Museum of Mining, Butte, Montana.

79. Model made for the Defendant, *Clark-Montana Realty Co. (Elm Orlu)* v. *Butte & Superior Mining Co. (Black Rock),* 1914–1919. Model on display at the World Museum of Mining, Butte, Montana.

80. "Unit Blocks for Mine Model," *Engineering and Mining Journal* 97 (May 1914): 1049–1050.

81. C. V. Hopkins, "United Verde Mine Model," *Mines and Minerals* 30 (March 1910): 501–502.

82. Not all skeleton models represented the surface. For instance, the West End model I describe in the next chapter showed the underground only.

83. Not all skeleton models followed this practice, however—the West End model I examine in the next chapter colored all of the workings with similar geology the same.

84. "Mine Model for Inclined Veins," *Engineering and Mining Journal* 90 (December 1910): 1243.

85. W. G. Zulch, "Stope Model," *Mines and Minerals* 32 (1912): 491.

86. Harrison and Zulch, "Court Maps and Models," 51–52.

87. "The Pennsylvania M. Co. vs. Grass Valley Exploration Co. Litigation," *Mining and Scientific Press* 83 (August 1901): 41, 44.

88. One vertical shaft on the model is marked "B. Rock Shaft." The Black Rock Shaft was on the Black Rock claim, owned by the Butte & Superior, and was involved in an apex lawsuit fight with the neighboring Elm Orlu. This model is held by the World Museum of Mining in Butte, Montana, which also holds another model labeled as having been made for the lawsuit between the Elm Orlu and Black Rock, though it is constructed very differently from this one.

89. For a diamond drill hole model using painted wires, see G. C. Bateman, "Diamond Drill Hole Model," *Engineering and Mining Journal* 95 (March 1913): 471.

90. "Graphic Indication of Drilling," *Engineering and Mining Journal* 89 (February 1910): 453.

91. W. St. J. Miller, "A Satisfactory Mine Model," *Mining and Scientific Press* 103 (1911): 779.

92. Benedict, *Red Metal*, 183–187.

93. C. H. Benedict, "Calumet & Hecla Reclamation Plant," *Proceedings of the Lake Superior Mining Institute* 24 (1925): 71, 78–79.

94. Benedict, *Red Metal*, 132–136.

95. Benedict, *Red Metal*, 191.

96. Many thanks to Jeremiah Mason, of Keweenaw National Historical Park, for locating the photograph used in figure 4.7 and providing a scan for me.

97. It is also notable that these three models are the earliest mention of glass plate models I have yet found. Hilary Bauerman, *A Descriptive Catalogue of the Geological, Mining, and Metallurgical Models in the Museum of Practical Geology* (London: George E. Eyre and William Spottiswoode, 1865), 102–105.

98. John R. Leifchild, *Cornwall: Its Mines and Miners; with Sketches of Scenery; Designed as a Popular Introduction to Metallic Mines* (London: Longman Brown Green and Longmans, 1855), 111–113; [quote on 112; emphasis in original].

99. William H. Storms, *Timbering and Mining: A Treatise on Practical American Methods*, 1st ed. (New York: McGraw-Hill, 1909), 243–244 [all quotes on 243–244].

100. Severy, "Making Mine Models."

101. Robert D. McCracken, *A History of Tonopah, Nevada* (Tonopah, NV: Nye County Press, 1990), 104.

102. Brough, *A Treatise on Mine-Surveying, 1888,* 264.

103. Brunton quoted in Rickard, *Interviews with Mining Engineers,* 81.

104. The model of the West End and Tonopah Extension mines on display at the

Tonopah Historic Mining Park was one such model, made for litigation that was settled before trial. For use of models in injunction hearings, see Reno Sales Oral History, Columbia Rare Book and Manuscript Library, p. 11.

105. Rickard, *Interviews with Mining Engineers*, 87–88.

106. "New Use for a Mine Model," *Mining and Scientific Press* 87 (October 1903): 247.

107. "New Use for a Mine Model," *Mining and Scientific Press* 87 (October 1903): 247.

108. "The Utah Consolidated Mining Company," *Engineering and Mining Journal* 82 (September 1906): 489.

109. Milton, "A Mine Model," 338, including quote.

110. Jordanova, "Material Models as Visual Culture," 449.

111. Mrs. Hugh (Marjorie) Brown, *Lady in Boomtown: Miners and Manners on the Nevada Frontier* (Palo Alto, CA: American West Publishing Co., 1968), 126.

5 ✕ Models and the Legal Landscape of Underground Mining

1. The West End's model survives in the W. M. Keck Museum, of the Mackay School of Mines at the University of Nevada–Reno. Many of the Jim Butler's maps made for the trial are held by the Tonopah Historic Mining Park (see esp. maps 791–815); a few Jim Butler maps and many of the West End maps were recorded by the US Bureau of Mines' microfilming efforts and are now available from the National Mine Map Repository (see, e.g., document numbers 400328, 400333, 400335, and 400336). The very complete transcripts, which newspapers at the time reported were constructed with special care, are located in the National Archives in Washington, DC. The case received some attention in the mining press of the time. Additionally, as complex as this case will undoubtedly seem, the issues are relatively simple compared with apex litigation in some districts, for example in Butte.

2. *Jim Butler Tonopah Mining Company* v. *West End Consolidated Mining Company*, 247 U.S. 450; 38 S. Ct. 574; 62 L. Ed. 1207 (1918).

3. On the history and importance of Tonopah, see Russell R. Elliott, *Nevada's Twentieth-Century Mining Boom: Tonopah, Goldfield, Ely* (Reno: University of Nevada Press, 1988); and McCracken, *A History of Tonopah, Nevada*. On the mining history of the town, see Jay A. Carpenter, Russell Richard Elliott, and Byrd Fanita Wall Sawyer, *The History of Fifty Years of Mining at Tonopah, 1900–1950*, University of Nevada Bulletin, 47, no. 1, Geology and Mining Series, No. 51 (Reno: Nevada Bureau of Mines and Geology, 1953), 55–58, 130–135; production statistics are on 149.

4. Carpenter et al., *Fifty Years of Mining at Tonopah*, 52, 66, 96.

5. Carpenter et al., *Fifty Years of Mining at Tonopah*, 52–53. Smith was old friends with some of the original locators, who had worked with him at his first borax operation, near Candelaria, Nevada, in the 1880s. See Hugh A. Shamberger, *The Story of*

Candelaria and its Neighbors: Columbus, Metallic City, Belleville, Marietta, Sodaville and Coaldale, Esmeralda and Mineral Counties, Nevada (Carson City, NV: Nevada Historical Press, 1978), 135–138.

6. Carpenter et al., *Fifty Years of Mining at Tonopah*, 66–67.

7. "Jim Butler Company Commences Suit Against West End Con," *Tonopah Miner*, April 11, 1914. The West End disputed the charges, even to the extent of disputing the tonnage and value of ore mined. The West End claimed, before the trial, that it had mined only 23,341 tons of ore with an aggregate net value of $204,300. "West End Stands Pat on Claim of Clear Cut Apex," *Tonopah Daily Bonanza*, July 7, 1914.

8. "West End Officers in Conference in New York," *Tonopah Daily Bonanza*, April 3, 1914; "Counter Proposition of Butler Reported by New York Message," *Tonopah Daily Bonanza*, April 4, 1914; "West End Stands Pat on Claim of Clear Cut Apex."

9. The term *extralateral rights* comes from famous American mining engineer Rossiter W. Raymond, who was an ardent foe of those rights. Rossiter W. Raymond, "The Law of the Apex," *Transactions of the American Institute of Mining Engineers* 12 (1884): 387–444; Rossiter W. Raymond, "The Law of the Apex—Appendix," *Transactions of the American Institute of Mining Engineers* 12 (1884): 677–688.

10. In an ideal rectangular mining claim, the end lines were the short sides of the rectangle, and the side lines were the long sides of the rectangle.

11. John L. Neff, "The Law of the Apex: A Continuing Enigma," *Rocky Mountain Mineral Law Institute* 18 (1973): 388–391; Agricola, *De Re Metallica*, esp. 80–86. Colby examined the existence of extralateral rights in other countries, past and present, in William E. Colby, "The Extralateral Right: Shall It Be Abolished?," *California Law Review* 4, no. 5 (1916): 361–388.

12. The terms *dip* and *strike* are the two measurements of a vein's location. Dip refers to the vein as it plunges into the earth, and strike describes the course of the vein, if a constant horizontal plane was assumed. Imagine a mountain with a vein embedded in it: if you sliced vertically through the mountain from top to bottom, perpendicular to the strike of the vein, you would see the dip of the vein revealed in the slice. Likewise, if you sliced off the top of the mountain perfectly horizontally, you would see the strike of the vein.

13. The 1872 mining law gave mining claimants "all veins, lodes, and ledges throughout their entire depth, the top or apex of which lies inside of such surface-lines extended downward vertically, although such veins, lodes, or ledges may so far depart from a perpendicular in their course downward as to extend outside the vertical side-lines of said surface locations." See "An Act to Promote the Development of the Mining Resources of the United States," *U.S. Statutes at Large* 17: 91–96; [quote on 91–92]. On the history of American mining law, see Bakken, *The Mining Law of 1872*; Charles W. Miller Jr., *Stake Your Claim! The Tale of America's Enduring Mining Laws* (Tucson, AZ: Westernlore Press, 1991); William T. Parry, *All Veins, Lodes, and Ledges Throughout Their Entire Depth: Geology and the Apex Law in Utah Mines* (Salt Lake City: University of Utah Press, 2004); and Bruce Alverson, "The Limits of

Power: Comstock Litigation, 1859–1864," *Nevada Historical Society Quarterly* 43, no. 1 (2000): 74–99. For a transnational perspective on mining law, see Barry Barton, "The History of Mining Law in the US, Canada, New Zealand and Australia, and the Right of Free Entry," in *International and Comparative Mineral Law and Policy,* ed. Elizabeth Bastida, Thomas Wälde, and Janeth Warden-Fernández (The Hague: Kluwer Law International, 2005), 643–660.

14. The two most important cases were *Iron Silver Mining Co.* v. *Cheesman* (1886) 116 U.S. 529, 6 S. Ct. 481, 29 L. Ed. 712; and *Iron Silver Mining Co.* v. *Murphy et al.* (1880), 3 F. 368. See also Raymond, "Law of the Apex."

15. For more biographical details on most of the experts and lawyers, see "Grand Galaxy of Talent in The Butler–West End Suit," *Tonopah Daily Bonanza,* December 17, 1914.

16. William E. Colby, "Curtis Holbrook Lindley," *California Law Review* 9 (1921): 87–99; also see John C. Lacy, "Pursuing Segments of the Glass Snake: The Battle for the Big Jim," in *History of Mining in Arizona,* ed. J. Michael Canty and Michael N. Greeley, vol. 2 (Tucson, AZ: Mining Club of the Southwest Foundation / the American Institute of Mining Engineers Tucson Section, 1991), 263–285.

17. Curtis H. Lindley, *A Treatise on the American Law Relating to Mines and Mineral Lands Within the Public Land States and Territories and Governing the Acquisition and Enjoyment of Mining Rights in Lands of the Public Domain,* 3rd ed. (San Francisco: Bancroft-Whitney, 1914) (hereafter called *Lindley on Mines*); earlier editions appeared in 1903 and 1897. Also see Lindley's 1913 article on secondary veins, which were a prominent part of the proceedings in the *Butler* v. *West End* case; Curtis H. Lindley, "A Problem in Extralateral Rights on Secondary Veins," *California Law Review* 1 (1913): 427–438.

18. Colby, "Curtis Holbrook Lindley," 91. Colby specifies these early works as Yale (1869), Copp's "U.S. Mineral Lands," Morrison's "Mining Rights," Weeks on Mineral Lands in 1877 and 1880, Wade's "American Mining Laws" (1889) and Sickel's "Mining Laws and Decisions" (1881).

19. Colby, "Curtis Holbrook Lindley," 92.

20. Horace V. Winchell, "Review of 'Lindley on Mines,'" *Economic Geology* 9, no. 6 (September 1914): 598–602.

21. Rossiter W. Raymond, "Review of 'A Treatise on the American Law Relating to Mines and Mineral Lands, Etc.,'" *Engineering and Mining Journal* 97 (April 1914): 817.

22. Colby, "Curtis Holbrook Lindley," 96–97; William E. Colby, "Curtis Holbrook Lindley," *Dictionary of American Biography* Base Set, American Council of Learned Societies, 1928–1936, reproduced in Biography Resource Center, Farmington Hills, MI: Thomson Gale, 2007.

23. Herbert Hoover, *The Memoirs of Herbert Hoover: Years of Adventure, 1874–1920* (New York: Macmillan, 1951), 27.

24. "Butler Geologist Presents Vivid Crayon Sketch," *Tonopah Daily Bonanza,*

December 16, 1914; "Jim Butler–West End Trial Concluded with Dispatch, Poetry, Banquet, and No Ill Feeling," *Tonopah Daily Bonanza,* December 22, 1914; "Muse Inspires Noted Jurist," *Tonopah Daily Bonanza,* December 23, 1914.

25. See William E. Colby, *Reminiscences: An Interview Conducted by Corinne L. Gilb* (Berkeley, CA: Regional Oral History Office, Bancroft Library, 1954), esp. 99–100; and Robert W. Righter, *The Battle over Hetch Hetchy: America's Most Controversial Dam and the Birth of Modern Environmentalism* (New York: Oxford University Press, 2005).

26. Don H. Sherwood, "The Extralateral Right: A Last Hurrah?," *Rocky Mountain Mineral Law Institute* 35 (1989): 12–2, n. 1. The similarly titled articles were Colby, "The Extralateral Right: Shall It Be Abolished?"; William E. Colby, "The Extralateral Right: Shall It Be Abolished? II: The Origin and Development of the Extralateral Right in the United States," *California Law Review* 4, no. 6 (1916): 437–464; William E. Colby, "The Extralateral Right: Shall It Be Abolished? III: The Federal Mining Act of 1872," *California Law Review* 5 (1916): 18–36; and William E. Colby, "The Extralateral Right: Shall It Be Abolished? IV: Conclusion," *California Law Review* 5 (1917): 303–330.

27. Stephen Cresswell, *Mormons and Cowboys, Moonshiners and Klansmen: Federal Law Enforcement in the South and West, 1870–1893* (Tuscaloosa: University of Alabama Press, 1991), 97–132. Members of the Mormon community charged that Dickson ignored Gentiles committing the same offenses. "Accused of Partiality," *Washington Post,* February 14, 1885. "Gentile" was the Mormon term for non-Mormons.

28. On the jars, see Cresswell, *Mormons and Cowboys,* 120; on the strike on the face, see "Assaulted by Mormons," *New York Times,* February 23, 1886; and Frank J. Cannon and Harvey J. O'Higgins, *Under the Prophet in Utah: The National Menace of a Political Priestcraft* (Boston: C. M. Clark, 1911), chap. 2.

29. These included *Lawson* v. *U.S. Mining Company, Mammoth Mining Company* v. *Grand Central Mining Company, Conkling Mining Company* v. *Silver King Mining Company, Thompson et al.* v. *Kearns et al., Niagara Mining Company* v. *Old Jordan Mining Company,* and *Eureka Hill Mining Company* v. *Bullion, Beck & Champion Mining Company.* List from "Grand Galaxy of Talent in The Butler–West End Suit."

30. Dickson left Salt Lake City in 1917 after the death of his wife, and lived in Los Angeles until his death in 1924. Nelson Knight, "This Old House: Dickson-Gardner-Wolfe Mansion," *The Capitol Hill Neighborhood Council Bulletin* [Salt Lake City, UT], no. 49 (May 2005): 2.

31. "Montana-Tonopah Holds Meeting in Reno," *Nevada State Journal,* July 2, 1905; for more on the Montana-Tonopah (which does not mention Dickson explicitly), see Carpenter et al., *Fifty Years of Mining at Tonopah,* 50–51, 64–65, 90–95.

32. "On August 1, 1901, the Cliffords conveyed by deed their right, title, and interest in the Wandering Boy, as well as in the Lucky Jim and Stone Cabin, to W. H. Dickson and A. C. Ellis, and on May 5, 1902, Dickson and Ellis conveyed the same to the complainant [the Tonopah and Salt Lake Mining Co.] herein." *Tonopah*

& *Salt Lake Mining Co.* v. *Tonopah Mining Co.*, 125 F. 400, 407 (1903). The other two lawsuits of the group had the same title and were reported as 125 F. 389 (1903) and 125 F. 408 (1903). The Stone Cabin and Wandering Boy claims were not part of the dispute between the Jim Butler and the West End, as the conflict was limited to the Eureka and Curtis claims of the Jim Butler and the West End claim of the West End Consolidated. For their ownership by the Jim Butler Company, see Carpenter et al., *Fifty Years of Mining at Tonopah*, 96.

33. "Judge Dickson Dies on Coast," *Deseret News*, January 18, 1924; "Grand Galaxy of Talent in The Butler–West End Suit." I would particularly like to thank Nelson Knight of the Utah Division of Historic Preservation for the *Deseret News* obituary and a photo of Dickson. Both Ellises appear frequently on reported court cases with Dickson. It is unclear which A. C. Ellis was involved with Dickson in the early Tonopah claims. A. C. Ellis Sr., died in 1912. J. P. O'Brien, ed., *History of the Bench and Bar of Nevada* (San Francisco: Bench and Bar, 1913), 79–80.

34. "Brown, Hugh Henry," in *Who Was Who in America*, vol. 1, 4th printing (Chicago: Marquis Who's Who, 1943), 149; O'Brien, *History of the Bench and Bar of Nevada*, 88; also see the well-known memoir of Brown's wife (Brown, *Lady in Boomtown*).

35. Brown worked for Campbell, Metson, & Campbell, which appears as part of the legal team in *Tonopah & Salt Lake Mining Co.* v. *Tonopah Mining Co. of Nevada*, 125 F. 389 (1903).

36. O'Brien, *History of the Bench and Bar of Nevada*, 96–97.

37. Harry Hunt Atkinson, *Tonopah and Reno Memories of a Nevada Attorney*, ed. Barbara C. Thornton (Reno: University of Nevada–Reno Library, Oral History Project, 1970), 21–23, 30–33, 43.

38. O'Brien, *History of the Bench and Bar of Nevada*, 95.

39. "Grand Galaxy of Talent in The Butler–West End Suit."

40. *Jim Butler Tonopah Mining Co.* v. *West End Consolidated Case File*, US Supreme Court Appellate Case Files, 25458, Number 249, Box 5000, Record Group 267, National Archives, Washington DC, 513–514; hereafter abbreviated as *Jim Butler* v. *West End* transcript.

41. "Grand Galaxy of Talent in The Butler–West End Suit." 3.

42. Dean Earl Bennett, *A Short History of the College of Mines and Earth Resources*, 2002; "PMG On Tour," *Time*, July 30, 1934; "Marginalia," *Time*, August 27, 1934.

43. The *Tonopah Daily Bonanza* spelled his name "Searles" throughout the trial, but the transcript of the case spells it correctly. For Searls' qualifications, see *Jim Butler* v. *West End* transcript, 856–857. Also see "Geology of the Tonopah Lodes Related by Experts," *Tonopah Daily Bonanza*, December 18, 1914, for Searls' testimony, in which he used a complex geological analogy, which Ramsey said was a distinct feature of his legal testimony. For Searls' career with Newmont, see Robert H. Ramsey, *Men and Mines of Newmont: A Fifty-Year History* (New York: Octagon Books, 1973),

39–45, 179. Searls was a top executive for Newmont between 1931 and 1966. Colby later tried and won several important extralateral rights cases for Searls concerning Newmont's Empire Star Mine in the Nevada City, California, mining district. Searls served as a member of the US Strategic Bombing Survey during World War II and was a member of Bernard Baruch's delegation to the United Nations' Atomic Energy Commission. Barton J. Bernstein, "The Quest for Security: American Foreign Policy and International Control of Atomic Energy, 1942–1946," *Journal of American History* 60, no. 4 (1974): 1032–1034. The Searls lawyers practiced almost exclusively in Nevada City, California, where Lindley was also extensively involved. The Searls' house and law library is now preserved as a research library as part of the state park system of California.

44. Andrew C. Lawson and Harry Fielding Reid, *The California Earthquake of April 18, 1906: Report of the State Earthquake Investigation Commission* (Washington, DC: Carnegie Institution of Washington, 1908).

45. Fred Searls Jr., "The Consultant," in *Andrew C. Lawson: Scientist, Teacher, Philosopher*, ed. Francis E. Vaughan (Glendale, CA: A. H. Clark, 1970), 205–211, lists cases for which Lawson testified. For his work for the MacNamara with Lindley against the West End, see *Jim Butler* v. *West End* transcript, 273.

46. For Winchell's glowing review of the third edition of *Lindley on Mines*, see Winchell, "Review of 'Lindley on Mines.'" Winchell was also president of the American Institute of Mining Engineers in 1919.

47. N. H. Winchell and H. V. Winchell, *The Iron Ores of Minnesota: Their Geology, Discovery, Development, Qualities, and Origin, and Comparison With Those of Other Iron Districts*, Geological and Natural History Survey of Minnesota Bulletin 6 (Minneapolis, MN: Harrison and Smith, State Printers, 1891).

48. Elizabeth Noble Shor, "Winchell, Horace Vaughn," in *American National Biography Online* (New York: Oxford University Press, 2000); "Grand Galaxy of Talent in The Butler–West End Suit"; Spence, *Mining Engineers and the American West*, 228–229.

49. He examined the gold mines in northern Korea, for example, which led to Western investment there. Spence, *Mining Engineers and the American West*, 285.

50. *Jim Butler* v. *West End* transcript, 174–176; "West End Lawyers in Apex Litigation Begin to Come," *Tonopah Daily Bonanza*, December 1, 1914. Juessen's post-trial career was not as successful. He took a management job at the New Almaden mercury mine in California, and a series of poor decisions that cost the company significant money led to the mine firing him. See Schneider, *Quicksilver*. Note: Pittsburg/h is inconsistently spelled throughout the documents. This book goes with the conventional way to spell it (with "h").

51. "Grand Galaxy of Talent in The Butler–West End Suit."

52. *Jim Butler* v. *West End* transcript, 213–214.

53. Carpenter et al., *Fifty Years of Mining at Tonopah*, 66–68.

54. For Chandler's reversal, see *Jim Butler* v. *West End* transcript, 235–310; ["Certainly he knows" (249)].

55. Chandler was apparently an excellent manager but died young, in late 1915. Carpenter et al., *Fifty Years of Mining at Tonopah,* 101, 104.

56. O'Brien, *History of the Bench and Bar of Nevada,* 84.

57. Carpenter et al., *Fifty Years of Mining at Tonopah,* 133.

58. *Jim Butler* v. *West End* transcript, b, h.

59. Lindley noted in his treatise that fairness would suggest that the company asserting an extralateral right should be allowed to open and close evidence if the extralateral right was the only issue. Lindley, *Lindley on Mines,* 2171.

60. This strategy may not have been apparent to all observers at first. The newspaper reporter noted that Lindley introduced four arguments: that it was not one vein on top of another, but just a mass of stringers; that the strike was confused for dip and vice versa; that the West End discovery shaft, on the north-dipping vein, did not convey rights to the south-dipping vein; and that the broken end line prohibited the exercise of extralateral rights. The argument that the structure was one vein instead of two makes the first point covered by the reporter necessary; but the journalist seemed to have missed the significance of the one-vein strategy. "Butler Presents Its Case," *Tonopah Daily Bonanza,* December 14, 1914.

61. The trial transcript is permeated throughout with such language; for a few examples of many, see *Jim Butler* v. *West End* transcript, 3–5, 18, 50.

62. *Jim Butler* v. *West End* transcript, 1–5.

63. *Jim Butler* v. *West End* transcript, 7–9.

64. *Jim Butler* v. *West End* transcript, 14–15.

65. The Butler's exhibits were numbered, and the West End's were lettered, so it would always be clear which litigant's displays were under discussion.

66. *Jim Butler* v. *West End* transcript, 9–27.

67. The light green color of the andesite cap was complemented by the dark green color of the other cap rock, described as Brougher dacite. Black or dark brown depicted rhyolite, blue stripes indicated faults, and gray (the color of the primer used on the model) did not depict any geology. The model maker painted on numbers and letters in black, and the whole structure was suspended above a claim map in black ink on light-colored paper. *Jim Butler* v. *West End* transcript, 11–13.

68. J. E. Spurr was the first geologist to report on Tonopah, and the Tonopah miners tended to use his geological nomenclature. For the most complete statement of his early work, see J. E. Spurr, *Geology of the Tonopah Mining District, Nevada,* USGS Professional Paper 42 (Washington, DC: GPO, 1905). J. A. Burgess took issue with Spurr's interpretation, and termed the rocks differently, and hypothesized geological origins different from those asserted by Spurr. See J. A. Burgess, "The Geology of the Producing Part of the Tonopah Mining District," *Economic Geology* 4 (1909): 681–712. Perhaps as a response to Burgess or as a consequence of further

development in the mines, Spurr modified some of his earlier findings, including changing some of the terms for the rocks, in J. E. Spurr, "Geology and Ore Deposits at Tonopah, Nev.," *Economic Geology* 10 (1915): 751–760. Augustus Locke, "The Geology of the Tonopah Mining District," *Transactions of the American Institute of Mining Engineers* 43 (1912): 157–166, summarized the dispute and tended to side with the interpretations of Burgess. Edson S. Bastin and Francis B. Laney, *The Genesis of the Ores at Tonopah, Nevada,* USGS Professional Paper 104 (Washington, DC: GPO, 1918) largely attempted to avoid the issue by basing their findings more on chemical analysis. Thomas B. Nolan, "The Underground Geology of the Tonopah Mining District, Nevada," *University of Nevada Bulletin* 29, no. 5 (1935): 1–49 is considered the last word on the subject. Nolan later headed the USGS.

69. Lindley squeezed this point out of Juessen in cross-examination; *Jim Butler* v. *West End* transcript, 206. The point is elaborated on 207–209, where Lindley went after the "abrupt change" on the model.

70. *Jim Butler* v. *West End* transcript, 15–16.

71. *Jim Butler* v. *West End* transcript, 177.

72. *Jim Butler* v. *West End* transcript, 179–180.

73. *Jim Butler* v. *West End* transcript, 220–227.

74. *Jim Butler* v. *West End* transcript, 230–232.

75. Lindley, *Lindley on Mines,* iv.

76. *Jim Butler* v. *West End* transcript, 488, 492–497; [quote on 496].

77. *Jim Butler* v. *West End* transcript, 497–498, 501–503.

78. *Jim Butler* v. *West End* transcript, 499.

79. *Jim Butler* v. *West End* transcript, 498–499.

80. "Mine Models in Tonopah Apex Litigation," *Engineering and Mining Journal* 99 (1915): 850. Averill's description of the model is *Jim Butler* v. *West End* transcript, 1138. Uren's brief description is *Jim Butler* v. *West End* transcript, 499.

81. Lindley clarified that Pierce was not a surveyor. *Jim Butler* v. *West End* transcript, 9.

82. *Jim Butler* v. *West End* transcript, 500–511.

83. *Jim Butler* v. *West End* transcript, 507–508.

84. *Jim Butler* v. *West End* transcript, 508–510; [quote on 510].

85. *Jim Butler* v. *West End* transcript, 510–511.

86. *Jim Butler* v. *West End* transcript, 509.

87. *Jim Butler* v. *West End* transcript, 1137. Averill's opinion is pages 1137–1158, but is also numbered 1–22, which are the numbers Averill uses when he refers to other pages in the opinion. The opinion was also printed in full, without the diagram, in "West End Wins Decision in Apex Case," *Tonopah Daily Bonanza,* April 30, 1915.

88. *Jim Butler* v. *West End* transcript, 1137, 1140. Averill refers to Lindley, *Lindley on Mines,* §312.

89. *Jim Butler* v. *West End* transcript, 1140.

90. *Jim Butler* v. *West End* transcript, 1142–1149.

91. *Jim Butler* v. *West End* transcript, 1146.

92. For example, *Jim Butler* v. *West End* transcript, 1142, 1145. Diagram is 1144. Diagram based on West End Model: 1143, lines 11–13.

93. *Jim Butler* v. *West End* transcript, 1147–1148.

94. *Jim Butler* v. *West End* transcript, 1149.

95. *Jim Butler* v. *West End* transcript, 1150.

96. One of the primary arguments made by the Jim Butler was that the West End vein had no terminal edge, so therefore it had no apex.

97. *Jim Butler* v. *West End* transcript, 1151–1154.

98. *Jim Butler* v. *West End* transcript, 1155.

99. *Jim Butler* v. *West End* transcript, 1155.

100. *Jim Butler* v. *West End* transcript, 1156.

101. *Jim Butler* v. *West End* transcript, 1156.

102. This doctrine was most fully and finally expressed by the US Supreme Court in a pair of cases, *Del Monte Mining and Milling Co.* v. *Last Chance Mining Co.*, 171 U.S. 55 (1898); and *Clark* v. *Fitzgerald,* 171 U.S. 92 (1898).

103. The other three were that the locator must make a discovery in place, give notice of his claim, and define his boundaries. The requirement for parallel end lines stemmed from the fourth requirement, which was that the boundary lines must conform to the requirements of the law. These requirements, defined elsewhere in the statute, mandated parallel end lines, unbroken lines, maximum sizes, and so on.

104. *Jim Butler* v. *West End* transcript, 1156–1157. Averill cited Lindley, *Lindley on Mines,* as well as Charles H. Shamel, *Mining, Mineral, and Geological Law: A Treatise on the Law of the United States* (New York: Hill Publishing Company, 1907).

105. Beatty quoted in *Jim Butler* v. *West End* transcript, 1157–1158.

106. See Lindley, *Lindley on Mines,* §312, 712. Lindley attributes the quote to Beatty's decision in *Gleeson* v. *Martin White M. Co.,* 13 Nev. 442 (1878), 459.

107. *Jim Butler* v. *West End* transcript, 1158.

108. *Jim Butler* v. *West End* transcript, 1142–1143, 1147.

109. *Jim Butler* v. *West End* transcript, 1140.

110. Brown letter to Curtis H. Lindley, April 30, 1915, in Nevada Mining District File 48400013, Nevada Bureau of Mines and Geology.

111. There was, and is, no intermediate court of appeals in the Nevada judicial system.

112. Nevada Supreme Court case no. 2,195; the case was reported as 39 Nev. 375; 158 Pac. 876, 1 A.L.R. 405.

113. 158 Pac. 876, 877.

114. 158 Pac. 876, 877

115. This section does not appear in the case as reported in 158 Pac. 876. See 39 Nev. 375. 1 A.L.R. 405 contains a small portion of this section. It is unknown why

this disparity between the reported versions exists, but it seems to be the only difference between the reports; [quotes in 39 Nev. at 386].

116. All quotes in this paragraph are from 39 Nev. at 387; "popular sense": 39 Nev. at 388.

117. Colby later argued "one of the most important mining cases that was ever tried in Nevada," which the attorney also described as "the most important . . . case I have ever been associated with" in front of Norcross. This was *Nevada Consolidated Mining Co. v. Nevada Consolidated Copper Mines*, 44 F.2d. 192 (1930); 64 F.2d. 440 (1933); cert. denied 290 U.S. 664 (1933). Colby, *Reminiscences*, 88–92; [quotes on 90].

118. 158 Pac. at 877.

119. 158 Pac. at 878. Also see *Del Monte Mining Co.*

120. 158 Pac. at 879.

121. 158 Pac. at 879.

122. 158 Pac. at 879.

123. 158 Pac. at 880.

124. 158 Pac. at 880; Norcross was quoting Lindley, *Lindley on Mines,* §305, which analyzes the legal conundrums posed by the Leadville cases.

125. 158 Pac. at 880.

126. Especially, according to Norcross, by Judge Hallett in *Iron Silver Mining Co. v. Murphy*, 3 F. 368 (1880); and by Justice Miller in *Stevens v. Williams,* Fed. Cas. 13,413 (1879).

127. 158 Pac. at 883.

128. McCarran was later a longtime US senator from Nevada and a close associate of Joseph McCarthy in the latter's "Red Scare" of the 1950s. See Jerome E. Edwards, *Pat McCarran: Political Boss of Nevada* (Reno: University of Nevada Press, 1982). See also 158 Pac. at 883. N.B.—the reported case discusses "C. J. Norcross" and "J. J. Coleman"—these initials apparently stand for chief justice and junior justice, respectively. See Dean Heller, ed., *Political History of Nevada, 1996,* 10th ed. (Carson City, NV: State Printing Office, 1997), 221, for the judges' full names.

129. 158 Pac. at 877, 878, 879.

130. The case is reported as 247 U.S. 450; 38 S. Ct. 574; 62 L. Ed. 1207. The *United States Reports* text includes a portion of the Butler's argumentation, in a summary form. The *United States Supreme Court Reports, Lawyer's Edition* includes a similar, though not word-for-word, rendition of the Butler's arguments, and also carries at least portions of the West End brief. The *Supreme Court Reporter* omits both, as does the version available online through FindLaw. I will refer to the paper edition of the *United States Reports* text except where necessary.

131. 62 L. Ed. at 1208–1209.

132. 247 U.S. at 453. Note here he is using a sense of the mining law that reflects the older sense, embodied in the 1866 law, that it is a matter of claim along a lode, rather than the square area of a claim, that is important.

133. 247 U.S. at 454–455.

134. 247 U.S. at 456–459 repeats 158 Pac. at 876–877.

135. 247 U.S. at 460.

136. 247 U.S. at 460.

137. The copper operators of Bisbee, Arizona, agreed amongst themselves to respect vertical sidelines, so as to avoid any type of apex litigation. Richard W. Graeme, "Bisbee, Arizona's Dowager Queen of Mining Camps: A Look at Her First 50 Years," in *History of Mining in Arizona,* ed. J. Michael Canty and Michael N. Greeley, vol. 1 (Tucson, AZ: Mining Club of the Southwest Foundation, 1987), 56.

138. Sherwood, "The Extralateral Right," 12.2–12.4.

139. 256 U.S. 18; 41 S. Ct. 426; 65 L. Ed. 811.

140. Dickson and Ellis appear on the portion of the brief included in 256 U.S. 18, as reported on Lexis-Nexis; Lindley appears along with Dickson on the brief for a case with the same name but reported differently; see *Silver King Coalition Mines Co. v. Conkling Mining Co.* (1921), 255 U.S. 151.

141. 256 U.S. 18.

142. 256 U.S. at 26.

143. 256 U.S. at 27.

144. Carpenter et al., *Fifty Years of Mining at Tonopah,* 98, 138.

145. Carpenter et al., *Fifty Years of Mining at Tonopah,* 106–114.

146. Carpenter et al., *Fifty Years of Mining at Tonopah,* 102; 3 B.T.A. 128, 130.

147. H. D. Budelman letter to Walter S. Palmer, August 30, 1939; L. W. Hartman letter to Walter S. Palmer, September 20, 1939; H. D. Budelman letter to Walter S. Palmer, September 22, 1939; all in the W. M. Keck Museum Administrative Files, Mackay School of Mines, University of Nevada–Reno. I am grateful to Rachel Dolbier, formerly of the Keck Museum, for finding these letters and sending copies to me.

6 ✕ Mine Models for Education and the Public

1. Harden and Harden, "The Construction of Maps in Relief," 301.

2. William Jones, *The Treasures of the Earth; or, Mines, Minerals, and Metals* (London: Frederick Warne, 1868), 192–193 [emphasis in original].

3. Jones, *The Treasures of the Earth,* 193–194 [emphasis in original].

4. J. C. Bartlett, "American Students of Mining in Germany," *Transactions of the American Institute of Mining Engineers* 5 (1877): 438.

5. Stoek, "Mine Models," 25–26.

6. W[alter] R. Crane, in discussion of Stoek, "Mine Models," 32.

7. Stoek, "Mine Models," 25.

8. Stoek, "Mine Models," 33.

9. John R. Chamberlin, in discussion of Stoek, "Mine Models," 32.

10. Thomas Egleston Papers (MS 0385), University Archives, Rare Book & Manuscript Library, Columbia University in the City of New York [hereafter Egleston

Papers]: Edward H. Russell to Thomas Eggleston [*sic*], December 14, 1894, folder "Russell, Edward H.," Box 1, Egleston Papers; Thomas Egleston to [Seth] Low, March 16, 1895, Egleston Papers; Egleston to Low, March 18, 1895, folder "Low, Seth (copies)," Box 1, Egleston Papers.

11. C. B. Parsons to H. S. Munroe, August 15, 1893; Parsons to Munroe, October 6, 1893, folder "Business & Professional Correspondence, 1890–99," Box 1, Henry Smith Munroe Collection (MS 0910), University Archives, Rare Book & Manuscript Library, Columbia University in the City of New York.

12. Stoek, "Mine Models," 26–30.

13. W[alter] R. Crane, "Concrete Mine Models," *Mines and Minerals* 27 (February 1907): 300.

14. Crane, "Concrete Mine Models," 300–302.

15. Walter R. Crane, *Ore Mining Methods,* 2nd ed. (New York: John Wiley and Sons, Inc., 1917).

16. W[alter] R. Crane, "Mining Methods Illustrated in Miniature," *Engineering and Mining Journal* 103 (1917): 563.

17. Stoek, "Mine Models," 33–35.

18. "Head-Frame Models Made of Paper," *Mines and Minerals* 30 (February 1910): 401–402.

19. Crane, in discussion of Stoek, "Mine Models," 33.

20. Sperr, in discussion of Stoek, "Mine Models," 35.

21. Thomas Egleston to Seth Low, President of Columbia College, February 13, 1890, folder "Low, Seth (copies)," Box 1, Egleston Papers.

22. Brough, *A Treatise on Mine-Surveying,* 1888, 264.

23. Daniels, in discussion of Stoek, "Mine Models," 31.

24. Rickard, *Interviews with Mining Engineers,* 335.

25. The early history of the Smithsonian and the United States National Museum (USNM) is fascinating. See particularly Pamela M. Henson, "'Objects of Curious Research': The History of Science and Technology at the Smithsonian," *Isis* 90, no. Supplement (1999): S249–S269; and Steven Lubar and Kathleen M. Kendrick, *Legacies: Collecting America's History at the Smithsonian* (Washington, DC: Smithsonian Institution Press, 2001); and William S. Walker, *A Living Exhibition: The Smithsonian and the Transformation of the Universal Museum* (Amherst: University of Massachusetts Press, 2013).

26. For Dewey's activities, see Box 20, RU 158, Smithsonian Institution Archives (hereafter SIA). On the centrality of the expositions to the USNM's collecting activities in the nineteenth century, see Henson, "Objects of Curious Research," S252–S254.

27. Lehman, "Topographical Models,"; Dewey, *A Preliminary Descriptive Catalogue,* 224; *Annual Report of the Board of Regents of the Smithsonian Institution . . . for the Year Ending June 30, 1884, Part II: Report of the United States National Museum* (Washington, DC: GPO, 1885), 261, 386.

28. W. H. Newhall to F. P. Dewey, July 1, 1886, RU 158, Box 20, folder 6, SIA.

29. Dewey, *A Preliminary Descriptive Catalogue of the Systematic Collections* 193; Accession 20762 records, RU 305, SIA [microfilm].

30. "Report on the Department of Economic Geology and Metallurgy, in the U.S. National Museum, 1889, by F. P. Dewey, Curator" (typescript) RU 158, Box 20, folder 9, SIA; Brunton, "Aspen Mountain."

31. *Annual Report of the Board of Regents of the Smithsonian Institution . . . for the Year Ending June 30, 1892: Report of the United States National Museum* (Washington, DC: GPO, 1892) (hereafter *USNM Annual Report for 1892*), 216–217.

32. R. A. F. Penrose, "Notes on the State Exhibits in the Mines and Mining Building at the World's Columbian Exposition, Chicago," *Journal of Geology* 1, no. 5 (1893): 457–470; detailed descriptions of many portions of the exhibits can be found in articles listed in David J. Bertuca, Donald K. Hartman, and Susan M. Neumeister, *The World's Columbian Exposition: A Centennial Bibliographic Guide* (Westport, CT: Greenwood Press, 1996), 126–129.

33. Duane A. Smith, Karen A. Vendl, and Mark A. Vendl, *Colorado Goes to the Fair: World's Columbian Exposition, Chicago, 1893* (Albuquerque: University of New Mexico Press, 2011).

34. For a complete description of the St. Louis World's Fair, see David Rowland Francis, *The Universal Exposition of 1904* (St. Louis, MO: Louisiana Purchase Exposition Company, 1913). Governmental reports, issued at the close of the Fair, generally offer comprehensive accounts of their exhibits. The classic historical study of world's fairs, including the St. Louis World's Fair, is Robert W. Rydell, *All the World's A Fair: Visions of Empire At American Industrial Expositions, 1876–1916* (Chicago: University of Chicago Press, 1984).

35. *Annual Report of the Board of Regents of the Smithsonian Institution . . . for the Year Ending June 30, 1905: Report of the U.S. National Museum* (Washington, DC: GPO, 1906) (hereafter *USNM Annual Report for 1905*), 62–63; RU 158, Box 61, Folder 9-10, SIA.

36. *USNM Annual Report for 1905*, 15–17.

37. *USNM Annual Report for 1905*, 6–8; *Report on the Progress and Condition of the United States National Museum for the Year Ending June 30, 1910* (Washington, DC: GPO, 1911), 13–15; *Report on the Progress and Condition of the United States National Museum for the Year Ending June 30, 1911* (Washington, DC: GPO, 1912), 12–18; *Report on the Progress and Condition of the United States National Museum for the Year Ending June 30, 1912* (Washington, DC: GPO, 1913), 12–14.

38. *Report on the Progress and Condition of the United States National Museum for the Year Ending June 30, 1913* (Washington, DC: GPO, 1914), 114; *Annual Report of the Board of Regents of the Smithsonian Institution . . . for the Year Ending June 30, 1912: Report of the United States National Museum* (Washington, DC: GPO, 1912) (hereafter *USNM Annual Report for 1912*), 59; "List of Materials Transferred to Mineral Technology, from the Dept. of Geology, June 1912," [typescript], RU 216, Box 2, SIA.

39. *Annual Report of the Board of Regents of the Smithsonian Institution . . . for the Year Ending June 30, 1914: Report of the United States National Museum* (Washington, DC: GPO, 1915) (hereafter *USNM Annual Report for 1914*), 136–137.

40. California state geologist to Chester Garfield Gilbert, [1916] in folder "Mineral Technology—Companies and Correspondence, 1916," drawer "Mineral Technology, A–C, 1 of 4 drawers," Mineral Technology Division Records, DWI.

41. *USNM Annual Report for 1914*, 57.

42. Gilbert to Dr. Frederick A. Goetz, March 29, 1916, folder "Mineral Techn.—Columbia University—1916–1917," drawer "Mineral Technology (A–C 1 of 4 drawers)," Mineral Technology Division Records, DWI. Similar letters were sent to other schools of mines, such as the University of Washington and the Michigan College of Mines. See Box 2, RU 297, SIA.

43. R. W. Raymond to C. G. Gilbert, April 6, 1916, folder "Mineral Techn.—Columbia University—1916–1917," drawer "Mineral Technology (A–C 1 of 4 drawers)," Mineral Technology Division Records, DWI.

44. The model was Mineral Technology Catalog #1, Accession 55791.

45. "Coal Mining at Louisiana Purchase Exposition: Description of Models Showing Works of Some of the Large Bituminous Coal Companies," *Mines and Minerals* 25 (September 1904): 82.

46. Gilbert memorandum to J. S. Goldsmith, March 9, 1915, folder "Mineral Technology—Companies and Correspondence, Jan 27 1915—Dec 7, 1915," drawer "Mineral Technology (A–C 1 of 4 drawers)," Mineral Technology Division Records, DWI.

47. *USNM Annual Report for 1914*, 135.

48. Carlton Jackson, *The Dreadful Month* (Bowling Green, OH: Bowling Green University Popular Press, 1982); Aldrich, *Safety First*, 73–74.

49. C. W. Mitman to A. Wetmore, March 2, 1943; Wetmore to Mitman, May 4, 1943; USNM shipping invoice to Corps of Engineers, August 19, 1943; accession 55791, reel 159, RU 305 [microfilm], SIA.

50. The company's name was inconsistently spelled, with and without the trailing "h."

51. The model was Mineral Technology catalog #269, Accession #56153. Though Walcott collected it in 1904, it was not accessioned until 1914, which represents the delay in forming the division.

52. *USNM Annual Report for 1914*, 135–136.

53. "Coal Mining at Louisiana Purchase Exposition," 81.

54. Gilbert letter to Johns-Manville Co., September 18, 1913, folder "Mineral Technology—Companies & Correspondence, 1913," drawer "Mineral Technology (A–C 1 of 4 drawers)," Mineral Technology Division Records, DWI.

55. Gilbert letter to Braun Corporation, August 13, 1913, folder "Mineral Technology—Companies & Correspondence, 1913," drawer "Mineral Technology" (A–C 1 of 4 drawers)," Mineral Technology Division Records, DWI

56. Gilbert letter to Deister Machine Co., August 19, 1913, folder "Mineral Technology—Companies & Correspondence, 1913," drawer "Mineral Technology (A–C 1 of 4 drawers)," Mineral Technology Division Records, DWI. Incidentally, Gilbert's pleas did not bear fruit—the Deister Machine Company politely but firmly refused to participate.

57. Gilbert letter to Galigher Machinery Company, September 5, 1913, folder "Mineral Technology—Companies & Correspondence, Jul 1913–Dec 1913," drawer "Mineral Technology (A–C 1 of 4 drawers)," Mineral Technology Division Records, DWI.

58. E.g., Gilbert to Walcott, April 15, 1914, giving a list of the major copper producers and Gilbert's preferences for models from each of them; RU 297, Box 2, SIA. Also Rathbun as intermediary, e.g., Gilbert to Rathbun, September 12, 1913; Gilbert to Rathbun, September 17, 1913; Rathbun to the Standard Oil Company, September 19, 1913; H. C. Folger Jr. to Rathbun, October 7, 1913; Gilbert to Folger, October 29, 1913; folder "Standard Oil Company, 1913," Box 2, RU 297, SIA.

59. [Chester G. Gilbert], "Memorandum to the Secretary," February 5, 1915, folder "Secretary 1910–1919," RU 297, Box 2, SIA.

60. Gilbert memorandum for the secretary, October 31, 1914; Gilbert to Dr. Walcott, November 24, 1914, folder "Secretary 1910–1919," RU 297, Box 2, SIA

61. On Jackling, see LeCain, *Mass Destruction*; and A.B. Parsons, *The Porphyry Coppers* (New York: American Institute of Mining and Metallurgical Engineers, 1933).

62. Gilbert to Walcott, July 8, 1914; Gilbert to Walcott, June 26, 1914; folder "Secretary, 1910–1919," RU 297, Box 2, SIA.

63. Gilbert to H.C. Goodrich, Chief Engineer, Utah Copper Company, June 19, 1917, folder "Utah Copper Company, 1914, 1916–1918," Box 2, RU 297, SIA, 1.

64. The panel was the cause of some delays, but the model was eventually put on display. Folder "Utah Copper Company, 1914, 1916–1918," Box 2, RU 297, SIA.

65. J. E. Burleson letter to Division of Mineral Technology, October 14, 1913, folder "Mineral Technology—Companies & Correspondence, 1913," drawer "Mineral Technology (A–C 1 of 4 drawers)," Mineral Technology Division Records, DWI.

66. Gilbert letter to B. H. Hite, April 18, 1916, folder "Mineral Technology—Companies and Correspondence, Jan 6 1916—Nov 15 1916," drawer "Mineral Technology (A–C 1 of 4 drawers)," Mineral Technology Division Records, DWI.

67. Gilbert letter to E. J. Pranke, September 22, 1917, folder "Mineral Technology—Companies & Correspondence," drawer "Mineral Technology (A–C 1 of 4 drawers)," Mineral Technology Division Records, DWI.

68. Gilbert memo to Ravenel, 11-22-1918, folder "William deC. Ravenel," box 2, RU 297, SIA, p. 4.

69. Gilbert letter to Benjamin B. Thayer, Anaconda Copper Mining Co., September 14, 1916, folder "Mineral Technology—Anaconda Copper Mining Co.," drawer "Mineral Technology (A–C 1 of 4 drawers)," Mineral Technology Division Records, DWI.

70. Arnon Gutfeld, "The Murder of Frank Little: Radical Labor Agitation in Butte,

Montana, 1917," *Labor History* 10, no. 2 (1969): 177–192; Arnon Gutfeld, "The Speculator Disaster in 1917: Labor Resurgence at Butte, Montana," *Arizona and the West*
11, no. 1 (1969): 27–38.

71. Gilbert letter to Thayer, October 23, 1917, folder "Mineral Technology
—Anaconda Copper Mining Co.," drawer "Mineral Technology (A–C 1 of 4 drawers)," Mineral Technology Division Records, DWI.

72. C. G. Gilbert, "Observations on the Use of Models in the Educational Work
of Museums," *Museum Work* 2, no. 3 (December 1919): 90.

73. C. G. Gilbert, "Observations on the Use of Models," 92.

74. C. G. Gilbert, "Observations on the Use of Models," 90–92.

75. See, e.g., Chester G. Gilbert, "Coal Products: An Object Lesson in Resource
Admininstration," *United States National Museum Bulletin* 102, no. 1 (1917): 1–16;
Joseph E. Pogue, "Sulphur: An Example of Industrial Independence," *United States
National Museum Bulletin* 102, no. 3 (1917): 1–10.

76. *Report on the Progress and Condition of the United States National Museum for
the Year Ending June 30, 1919* (Washington, DC: GPO, 1920), 18, 37. An illustrative example: During the annual shutdown of the museum's heating and electricity
plant for the summer months, intended to permit repairs and allow the maintenance employees to take their annual leave, the entire force of assistant engineers
and almost all of the firemen and laborers found much more lucrative jobs elsewhere. Pay raises of $10 to $15 per month were insufficient to coax them back, so
the museum's physical plant was staffed by new employees when the boilers were
restarted in September 1918. *Annual Report of the Board of Regents of the Smithsonian
Institution . . . for the Year Ending June 30, 1919: Report of the United States National
Museum* (Washington, DC: GPO, 1919) (hereafter *USNM Annual Report for 1919*), 16.

77. Folder "William deC. Ravenel, 1919," Box 2, RU 297, SIA.

78. Mitman instead worked to create a separate Museum of Engineering and
Industry. Arthur P. Molella, "The Museum That Might Have Been: The Smithsonian's National Museum of Engineering and Industry," *Technology and Culture* 32, no.
2, pt. 1 (April 1991): 237–263.

79. "Report on the Division of Engineering, 1931–32 [typescript]," folder 15, box
63, RU 158, SIA, p. 27.

80. Gilbert letter to E. J. Pranke, September 22, 1917, folder "Mineral Technology
—Companies & Correspondence," drawer "Mineral Technology (A–C 1 of 4 drawers)," Mineral Technology Division Records, DWI.

Conclusion

1. John A. Rockwell, *A Compilation of Spanish and Mexican Law, in Relation to
Mines, and Titles to Real Estate, in Force in California, Texas, and New Mexico,* vol. 1
(New York: John S. Voorhies, 1851), 65.

2. "An Act for the Better Regulation and Ventilation of Mines," 855; "An Act Providing for the Health and Safety of Persons Employed in Coal Mines," 3.

3. "Colliery Notes," *Engineering and Mining Journal* 91 (1911): 431.

4. Robert P. Wolensky, Kenneth C. Wolensky, and Nicole H. Wolensky, *The Knox Mine Disaster, January 22, 1959: The Final Years of the Northern Anthracite Industry and the Effort to Rebuild a Regional Economy* (Harrisburg, PA: Pennsylvania Historical / Museum Commission, 1999); also Robert P. Wolensky, Kenneth C. Wolensky, and Nicole H. Wolensky, *Voices of the Knox Mine Disaster: Stories, Remembrances, and Reflections of the Anthracite Coal Industry's Last Major Catastrophe, January 22, 1959* (Harrisburg, PA: Pennsylvania Historical / Museum Commission, 2005).

5. David DeKok, *Fire Underground: The Ongoing Tragedy of the Centralia Mine Fire*, Rev. ed. (Guilford, CT: Globe Pequot Press, 2009).

6. Ralph H. Whaite, *Microfilming Maps of Abandoned Anthracite Mines: Mines of the Eastern Middle Field*, Bureau of Mines Information Circular 8274 ([Washington, DC]: United States Department of the Interior, Bureau of Mines, 1965), 1, 3.

7. Whaite, *Microfilming Maps of Abandoned Anthracite Mines*, 3.

8. Whaite, *Microfilming Maps of Abandoned Anthracite Mines*, 11.

9. Whaite, *Microfilming Maps of Abandoned Anthracite Mines*, 5.

10. Whaite, *Microfilming Maps of Abandoned Anthracite Mines*, 12–14.

11. Whaite, *Microfilming Maps of Abandoned Anthracite Mines*, 3, 11–12.

12. Whaite, *Microfilming Maps of Abandoned Anthracite Mines*, 15.

13. Walter L. Eaton and George B. Gait, *Microfilming Maps of Abandoned Anthracite Mines: Mines in the Wyoming Basin, Northern Anthracite Field*, Bureau of Mines Information Circular 8379 ([Washington, DC]: U.S. Department of the Interior, Bureau of Mines, 1968), 10.

14. George B. Gait, *Microfilming Maps of Abandoned Anthracite Mines: Mines in the Lackawanna Basin, Northern Anthracite Field*, Bureau of Mines Information Circular 8453 ([Washington, DC]: U.S. Department of the Interior, Bureau of Mines, 1970), 7; George B. Gait, *Microfilming Maps of Abandoned Anthracite Mines: Mines in the Western Middle Anthracite Field*, Bureau of Mines Information Circular 8519 ([Washington]: U.S. Department of the Interior, Bureau of Mines, 1971), 8.

15. George B. Gait, *Microfilming Maps of Abandoned Anthracite Mines: Mines in the Southern Anthracite Field*, Bureau of Mines Information Circular 8779 ([Washington]: U.S. Department of the Interior, Bureau of Mines, 1978), 9.

16. Curtis D. Edgerton, *The Mine Map Repository: A Source of Mine Map Data*, Bureau of Mines Information Circular 8657 ([Washington, DC]: US Bureau of Mines, 1974), 1–2. A third repository later opened in Spokane, Washington State, and there was a small repository active in Juneau, Alaska, at one time.

17. Edgerton, *The Mine Map Repository*, 2.

18. "Federal Coal Mine Health and Safety Act of 1969," P.L. 91-173, 83 Stat. 742. This has been amended several times since 1969. Current regulations regarding underground maps are found under 30 CFR 75.1200.

19. Edgerton, *The Mine Map Repository*, 2–8.

20. David C. Uhrin, *Correlating Microfilm Mine Maps with Topographic Maps*,

Bureau of Mines Information Circular 8762 ([Washington, DC]: US Department of the Interior, Bureau of Mines, 1977).

21. Office of Surface Mining Reclamation and Enforcement, *Mine Map Repositories: A Source of Mine Map Data,* (US Department of the Interior, Office of Surface Mining Reclamation and Enforcement, Program Information Development, April 1989), 1, 4–5.

22. John N. Murphy, "Update on the Continuing Functions of the Former US Bureau of Mines," *Mining Engineering* 49 (June 1997): 88. The Wilkes-Barre repository was moved to Pittsburgh in 2011.

23. "MSHA Finds New Map of Flooded Mine," *Coal Age* 107 (October 2002): 6.

24. "MSHA Announces Grants to Digitize Abandoned Mine Maps," *Pit & Quarry* 96 (February 2004): 55; Dibya Sarkar, "States to Make Digital Maps for Mines," *Federal Computer Week* 18 (January 2004): 14; "Digitized Mine Maps," US Department of Labor, http://www.msha.gov/minemapping/minemapping.asp; accessed 12-10-2012; author e-mail with Paul Coyle, November 30, 2012.

BIBLIOGRAPHY

Archives and Museum Collections

Anaconda Geological Collection. American Heritage Center, University of Wyoming, Laramie.

Brunton: David W. Brunton Collection. Western History Collection, Denver Public Library.

Calumet and Hecla Collection. Keweenaw National Historical Park, Calumet, MI.

Calumet and Hecla Drawing Collection (MS-005). Copper Country Historical Collections and Michigan Technological University Archives, Houghton, MI.

Calumet and Hecla Unprocessed Drawings. Copper Country Historical Collections and Michigan Technological University Archives, Houghton, MI.

Coxe: Eckley Brinton Coxe Manuscripts. Lehigh University Special Collections, Bethlehem, PA.

Coxe Family Mining Papers. Historical Society of Pennsylvania, Philadelphia.

Coxe Map Collection. Division of Work and Industry. National Museum of American History, Smithsonian Institution, Washington, DC.

Egleston: Thomas Egleston Papers. Columbia University Rare Book and Manuscript Library, New York.

Hoffman, John, Curatorial Files. Division of Work and Industry, National Museum of American History, Smithsonian Institution, Washington, DC.

Mineral Technology Divisional Records. Division of Work and Industry, National Museum of American History, Smithsonian Institution, Washington, DC.

Mining Engineering Project Manuscripts. Columbia University Oral History Research Office, New York.

Munroe, Henry Smith, Collection. Columbia University Rare Book and Manuscript Library, New York.

National Mine Map Repository. Office of Surface Mining, U.S. Department of the Interior, Washington, DC.

Nevada Mining District Files (http://www.nbmg.unr.edu/mdfiles/mdfiles.htm) Nevada Bureau of Mines and Geology, Reno, NV

Quincy Mining Co. Collection. Keweenaw National Historical Park, Calumet, MI.

Quincy Mining Company Collection (MS-001). Copper Country Historical Collections and Michigan Technological University Archives, Houghton, MI.

Quincy Mining Company Engineering Drawings Collection (MS-012). Copper

Country Historical Collections and Michigan Technological University Archives, Houghton, MI.

Student thesis manuscripts. Lehigh University Special Collections, Bethlehem, PA.

Tonopah Historic Mining Park Map Collection. Tonopah Historic Mining Park, Tonopah, NV.

United States National Museum Accession Records (RU 305) [microfilm]. Smithsonian Institution Archives, Washington, DC.

United States National Museum Curators' Annual Reports (RU 158). Smithsonian Institution Archives, Washington, DC.

United States National Museum Division of Engineering Records (RU 297). Smithsonian Institution Archives, Washington, DC.

United States National Museum Divisions of Mineral and Mechanical Technology Collection Records (—RU 216). Smithsonian Institution Archives, Washington, DC.

U.S. Supreme Court Appellate Case Files (RG 267). National Archives, Washington DC.

U.S. Supreme Court Records and Briefs, 1832–1978. Gale [online database].

Winchell: Newton Horace Winchell and Family Collection. Minnesota Historical Society, St. Paul, MN.

W. M. Keck Museum Administrative Files. Mackay School of Mines, University of Nevada–Reno.

World Museum of Mining, Butte, MT.

Other Published Sources

Abbott, Andrew. *The System of Professions: An Essay on the Division of Expert Labor.* Chicago: University of Chicago Press, 1988.

Abel, Sir Frederick Augustus. *Mining Accidents and Their Prevention: With Discussion by Leading Experts.* New York: The Scientific Publishing Company, 1889.

"An Act for the Better Regulation and Ventilation of Mines, and for the Protection of the Lives of the Miners in the County of Schuylkill." In *Laws of the General Assembly of the State of Pennsylvania Passed at the Session of 1869,* 852–856. Harrisburg, PA: B. Singerly, State Printer, 1869.

"An Act Providing for the Health and Safety of Persons Employed in Coal Mines." In *Laws of the General Assembly of the State of Pennsylvania Passed at the Session of 1870,* 3–12. Harrisburg, PA: B. Singerly, State Printer, 1870.

"An Act to Promote the Development of the Mining Resources of the United States," *U.S. Statutes at Large* 17: 91–96.

Agricola, Georgius. *De Re Metallica.* Translated by Herbert Clark Hoover and Lou Henry Hoover. London: The Mining Magazine, 1912. First published 1556.

Aldrich, Mark. *Safety First: Technology, Labor, and Business in the Building of American Work Safety, 1870–1939.* Baltimore: Johns Hopkins University Press, 1997.

Alverson, Bruce. "The Limits of Power: Comstock Litigation, 1859–1864." *Nevada Historical Society Quarterly* 43, no. 1 (2000): 74–99.

"An Interesting Mine Model." *Mining and Scientific Press* 29 (August 1874): 136.

Andrews, Thomas G. *Killing for Coal: America's Deadliest Labor War.* Cambridge, MA: Harvard University Press, 2008.

Angel, Myron, ed. *History of Nevada, with Illustrations and Biographical Sketches of Its Prominent Men and Pioneers.* Oakland, CA: Thompson and West, 1881.

Annual Report of the Board of Regents of the Smithsonian Institution . . . for the Year Ending June 30, 1884: Part II [Report of the United States National Museum] (Washington, DC: GPO, 1885).

Annual Report of the Board of Regents of the Smithsonian Institution . . . for the Year Ending June 30, 1892: Report of the United States National Museum (Washington, DC: GPO, 1893).

Annual Report of the Board of Regents of the Smithsonian Institution . . . for the Year Ending June 30, 1905: Report of the U.S. National Museum. Washington, DC: GPO, 1906.

Annual Report of the Board of Regents of the Smithsonian Institution . . . for the Year Ending June 30, 1919: Report of the United States National Museum (Washington, DC: GPO, 1920).

Arnold, Philip A. "Surveying and Mapping." *Coal Age* 12 (1917): 1109–1110.

Ashburner, Charles A. *First Report of Progress in the Anthracite Coal Region: The Geology of the Panther Creek Basin or Eastern End of the Southern Field.* Second Geological Survey of Pennsylvania, Report AA. Harrisburg, PA: Board of Commissioners for the Second Geological Survey, 1883.

———. "New Method of Mapping the Anthracite Coal-Fields of Pennsylvania." *Transactions of the American Institute of Mining Engineers* 9 (1881): 506–518.

Atkinson, Harry Hunt. *Tonopah and Reno Memories of a Nevada Attorney.* Edited by Barbara C. Thornton. Reno: University of Nevada–Reno Library, Oral History Project, 1970.

Averill, Volney. "Details and Cost of a Mine Model." *Engineering and Mining Journal* 108 (1919): 824–825.

Bailey, Lynn R. *Supplying the Mining World: the Mining Equipment Manufacturers of San Francisco, 1850–1900.* Tucson, AZ: Westernlore Press, 1996.

Bakken, Gordon Morris. *The Mining Law of 1872: Past, Politics, and Prospects.* Albuquerque: University of New Mexico Press, 2008.

Barnes, P. "Note upon the 'Blue' Process of Copying Tracings Etc." *Transactions of the American Institute of Mining Engineers* 6 (1879): 197–198.

———. "Note upon the So-Called Blue Process of Copying Tracings, Etc." *Engineering News* (October 1878).

Bartlett, J. C. "American Students of Mining in Germany." *Transactions of the American Institute of Mining Engineers* 5 (1877): 431–447.

Barton, Barry. "The History of Mining Law in the US, Canada, New Zealand and

Australia, and the Right of Free Entry." In *International and Comparative Mineral Law and Policy,* edited by Elizabeth Bastida, Thomas Wälde, and Janeth Warden-Fernández, 643–660. The Hague: Kluwer Law International, 2005.

Bastin, Edson S., and Francis B. Laney. *The Genesis of the Ores at Tonopah, Nevada.* U.S. Geological Survey Professional Paper 104. Washington, DC: GPO, 1918.

Bateman, G. C. "Diamond Drill Hole Model." *Engineering and Mining Journal* 95 (March 1913): 471.

Bauerman, Hilary. *A Descriptive Catalogue of the Geological, Mining, and Metallurgical Models in the Museum of Practical Geology.* London: George E. Eyre and William Spottiswoode, 1865.

Beard, James T. *Mine Examination Questions and Answers.* 1st ed. New York: McGraw-Hill, 1923.

Becker, George F. *Geology of the Quicksilver Deposits of the Pacific Slope.* U.S. Geological Survey Monograph 13. Washington, DC: GPO, 1888.

Benedict, C. H[arry]. "Calumet & Hecla Reclamation Plant." *Proceedings of the Lake Superior Mining Institute* 24 (1925): 68–88.

———. *Red Metal: The Calumet and Hecla Story.* Ann Arbor: University of Michigan Press, 1952.

Bernstein, Barton J. "The Quest for Security: American Foreign Policy and International Control of Atomic Energy, 1942–1946." *Journal of American History* 60, no. 4 (1974): 1003–1044.

Bertuca, David J., Donald K. Hartman, and Susan M. Neumeister. *The World's Columbian Exposition: A Centennial Bibliographic Guide.* Westport, CT: Greenwood Press, 1996.

Bewick, Thomas J. "Memoir of Mr. Thomas Sopwith." *Transactions of the North of England Institute of Mining and Mechanical Engineers* 29 (1880): 105–111.

Blackburn, Renee M. "Preserving and Interpreting the Mining Company Office: Landscape, Space, and Technological Change in the Management of the Copper Industry." M.S. Thesis, Michigan Technological University, 2011.

Bohning, James J. "Angel of the Anthracite: The Philanthropic Legacy of Sophia Georgina Coxe." *Canal History and Technology Proceedings* 24 (2005): 150–182.

———. "Chemistry, Coal and Culture: The Library of Eckley Brinton Coxe." Paper delivered at the National Meeting of the American Chemical Society, Chicago, August 2001.

Botsford, H. L. "An Index System for Maps." *Engineering and Mining Journal* 94 (1912): 638.

Bradsby, H. C., ed. *History of Luzerne County, Pennsylvania, with Biographical Selections.* Chicago: S. B. Nelson & Co., 1893.

Brightly, Frank F., ed. *Brightly's Purdon's Digest: A Digest of the Statute Law of the State of Pennsylvania from the Year 1700 to 1894.* 12th ed., vol. 2. Philadelphia: Kay and Brother, 1894.

Brough, Bennett H. *A Treatise on Mine-Surveying.* London: Charles Griffin and Company; Philadelphia: J. B. Lippincott, 1888.

———. *A Treatise on Mine-Surveying.* 12th ed. London: Charles Griffin and Company, 1906.

Brough, Bennett H., and Harry Dean. *A Treatise on Mine-Surveying.* 17th ed., rev. by Henry Louis. London: C. Griffin and Company, 1926.

Brown, Elspeth H. *The Corporate Eye: Photography and the Rationalization of American Commercial Culture, 1884–1929.* Baltimore: Johns Hopkins University Press, 2005.

Brown, George E. *Ferric and Heliographic Processes: A Handbook for Photographers, Draughtsmen, and Sun Printers.* New York: Tennant & Ward, 1900.

Brown, John K. "When Machines Became Gray and Drawings Black and White: William Sellers and the Rationalization of Mechanical Engineering." *IA: The Journal of the Society for Industrial Archeology* 25, no. 2 (1999): 29–54.

Brown, Marjorie (Mrs. Hugh). *Lady in Boomtown: Miners and Manners on the Nevada Frontier.* Palo Alto, CA: American West Publishing Co., 1968.

Browne, J. Ross. *Letter from the Secretary of the Treasury Transmitting A Report upon the Mineral Resources of the States and Territories West of the Rocky Mountains.* 39th Cong, 2d Sess., House of Representatives Ex. Doc. No. 29. Washington, DC: GPO, 1867.

Brunton, D[avid] W. "Aspen Mountain: Its Ores and their Mode of Occurrence." *Engineering and Mining Journal* 46 (1888): 22–23, 42–45.

———. "Geological Mine-Maps and Sections." *Transactions of the American Institute of Mining Engineers* 36 (1906): 508–540.

———. "Modern Progress in Mining and Metallurgy in the Western United States." *Transactions of the American Institute of Mining Engineers* 40 (1910): 543–561.

———. "A New System of Ore-Sampling." *Transactions of the American Institute of Mining Engineers* 13 (1885): 639–645.

"Brunton Awarded First Mining Medal." *Mining and Metallurgy* 8 (February 1927): 82–83.

Bucky, Philip B. "Use of Models for the Study of Mining Problems." *American Institute of Mining and Metallurgical Engineers Technical Publication,* no. 425 (1931): 3–28.

Burgess, J. A. "The Geology of the Producing Part of the Tonopah Mining District." *Economic Geology* 4 (1909): 681–712.

Byler, E. A., and Lee W. Davis. "Topographic Model of Cripple Creek District." *Mining and Scientific Press* 107 (1913): 144.

Campbell, C. M. "Methods of Mining the Granby Orebodies." *Engineering and Mining Journal* 87 (January 1909): 252–256.

Cannon, Frank J., and Harvey J. O'Higgins. *Under the Prophet in Utah: The National Menace of a Political Priestcraft.* Boston: C. M. Clark, 1911. http://www.gutenberg.org/etext/7066

Carpenter, Jay A., Russell Richard Elliott, and Byrd Fanita Wall Sawyer. *The History of Fifty Years of Mining at Tonopah, 1900–1950.* University of Nevada Bulletin, 47, no. 1, Geology and Mining Series, No. 51. Reno: Nevada Bureau of Mines and Geology, 1953.

Chadarevian, Soraya de, and Nick Hopwood, eds. *Models: The Third Dimension of Science.* Stanford, CA: Stanford University Press, 2004.

Chance, H. M. *Report on the Mining Methods and Appliances Used in the Anthracite Coal Fields.* Harrisburg, PA: Second Geological Survey of Pennsylvania, 1883.

Clemmer, Gregg S. *American Miners' Carbide Lamps: A Collector's Guide to American Carbide Mine Lighting.* Tucson, AZ: American Miners' Carbide Lamps: A Collector's Guide to American Carbide Mine Lighting, 1987.

"Coal Mining at Louisiana Purchase Exposition: Description of Models Showing Works of Some of the Large Bituminous Coal Companies." *Mines and Minerals* (September 1904): 81–84.

Colby, William E. "Curtis Holbrook Lindley." *California Law Review* 9 (1921): 87–99.

———. "The Extralateral Right: Shall It Be Abolished?" *California Law Review* 4, no. 5 (1916): 361–388.

———. "The Extralateral Right: Shall It Be Abolished? II: The Origin and Development of the Extralateral Right in the United States." *California Law Review* 4, no. 6 (1916): 437–464.

———. "The Extralateral Right: Shall It Be Abolished? III: The Federal Mining Act of 1872." *California Law Review* 5 (1916): 18–36.

———. "The Extralateral Right: Shall It Be Abolished? IV: Conclusion." *California Law Review* 5 (1917): 303–330.

———. *Reminiscences: An Interview Conducted by Corinne L. Gilb.* Berkeley, CA: Regional Oral History Office, Bancroft Library, 1954.

"Collapsible Geological Model." *Engineering and Mining Journal* 99 (1915): 532–533.

"Colliery Notes." *Engineering and Mining Journal* 91 (1911): 431.

Cook, James W. "Seeing the Visual in U.S. History." *Journal of American History* 95, no. 2 (2008): 432–441.

Coxe, Eckley B. "A Furnace with Automatic Stoker, Travelling Grate, and Variable Blast, Intended Especially for Burning Small Anthracite Coals." *Transactions of the American Institute of Mining Engineers* 22 (1894): 581–606.

———. "Improved Method of Measuring in Mine Surveys." *Transactions of the American Institute of Mining Engineers* 2 (1875): 219–224.

———. "The Iron Breaker at Drifton, With a Description of Some of the Machinery Used for Handling and Preparing Coal at the Cross Creek Collieries." *Transactions of the American Institute of Mining Engineers* 19 (1891): 398–474.

———. *Mining Legislation.* Author's offprint, Lehigh University Special Collections. Philadelphia: A Paper Read at the General Meeting of the American Social Science Association, October 1870.

———. "Presidential Address, Proceedings of the Lake George and Lake Champlain Meeting, October 1878." *Transactions of the American Institute of Mining Engineers* 7 (1879): 103–114.

———. "Remarks on the Use of the Plummet-Lamp in Underground Surveying." *Transactions of the American Institute of Mining Engineers* 1 (1873): 378–379.

———. "Secondary Technical Education." *Transactions of the American Institute of Mining Engineers* 7 (1879): 217–226.

———. "Some Thoughts upon the Economical Production of Steam, With Special Reference to the Use of Cheap Fuel." *Transactions of the New England Cotton Manufacturers' Association* 58 (1895): 133–210.

———. "Technical Education." *Transactions of the American Society of Mechanical Engineers* 15 (1894): 655–668.

Coxe, Eckley B., Heber S. Thompson, and William Griffith. *Report of the Commission Appointed to Investigate the Waste of Coal Mining, With the View to the Utilizing of the Waste.* Philadelphia: Allen, Lane & Scott's Printing House, May 1893.

Craig, W. H. "Geotechnical Centrifuges: Past, Present, and Future." In *Geotechnical Centrifuge Technology*, edited by R. N. Taylor, 1–19. Glasgow: Blackie Academic / Chapman and Hall, 1995.

Crane, W[alter] R. "Concrete Mine Models." *Mines and Minerals* 27 (February 1907): 300–302.

———. "Mining Methods Illustrated in Miniature." *Engineering and Mining Journal* 103 (1917): 563–566.

———. *Ore Mining Methods.* 2nd ed. New York: John Wiley and Sons, 1917.

Cresswell, Stephen. *Mormons and Cowboys, Moonshiners and Klansmen: Federal Law Enforcement in the South and West, 1870–1893.* Tuscaloosa: University of Alabama Press, 1991.

Cronin, Marionne. "Northern Visions: Aerial Surveying and the Canadian Mining Industry, 1919–1928." *Technology and Culture* 48 (April 2007): 303–330.

"David William Brunton." *Mining and Metallurgy* 9 (January 1928): 37.

DeKok, David. *Fire Underground: The Ongoing Tragedy of the Centralia Mine Fire.* Revised edition. Guilford, CT: Globe Pequot Press, 2009.

Dewey, Frederic P. *A Preliminary Descriptive Catalogue of the Systematic Collections in Economic Geology and Metallurgy in the United States National Museum.* Washington, DC: GPO, 1891.

Dikovitskaya, Margaret. *Visual Culture: The Study of the Visual after the Cultural Turn.* Cambridge, MA: MIT Press, 2005.

Dunshee, B. H. "Shaft Sinking in Butte, Montana." *Mines and Minerals* 27 (1907): 262–263.

Eaton, Walter L., and George B. Gait. *Microfilming Maps of Abandoned Anthracite Mines: Mines in the Wyoming Basin, Northern Anthracite Field.* Bureau of Mines Information Circular 8379. [Washington, DC]: US Department of the Interior, Bureau of Mines, 1968.

Edgerton, Curtis D. *The Mine Map Repository: A Source of Mine Map Data.* Bureau of Mines Information Circular 8657. [Washington, DC]: US Bureau of Mines, 1974.

Edwards, Jerome E. *Pat McCarran: Political Boss of Nevada.* Reno: University of Nevada Press, 1982.

Elliott, Russell R. *Nevada's Twentieth-Century Mining Boom: Tonopah, Goldfield, Ely.* Reno: University of Nevada Press, 1988.

Emmons, David M. *The Butte Irish: Class and Ethnicity in an American Mining Town, 1875–1925.* Urbana: University of Illinois Press, 1989.

Emmons, Samuel Franklin. *Geology and Mining Industry of Leadville, Colorado.* US Geological Survey Monograph 12. Washington, DC: GPO, 1886.

"The *Eureka-Richmond* Case." *Engineering and Mining Journal* 24 (September 1877): 181–182.

Evans, W. I. "A New Method of Making Mine Models." *Engineering and Mining Journal* 58 (1894): 293.

Farrel, J. H. "Geological Mapping of Mine Workings." *Engineering and Mining Journal* 86 (1908): 385.

Fay, Albert H. *Coal-Mine Fatalities in the United States, 1870–1914.* US Bureau of Mines Bulletin 115. Washington: GPO, 1916.

Ferguson, Eugene S. *Engineering and the Mind's Eye.* Cambridge, MA: MIT Press, 1992.

———. "The Mind's Eye: Nonverbal Thought in Technology." *Science* 197 (August 1977): 827–836.

Fickett, H. L. "Mine Maps vs. Mine Sketches." *Coal Age* 8 (1915): 740.

Foote, Mary Hallock. *A Victorian Gentlewoman in the Far West: the Reminiscences of Mary Hallock Foote.* Edited by Rodman W. Paul. San Marino, CA: Huntington Library, 1972.

Foster, Rufus J. "Protect Mine Maps and Plans." *Coal Age* 10 (1916): 376.

Francis, David Rowland. *The Universal Exposition of 1904.* St. Louis: Louisiana Purchase Exposition Company, 1913.

Francoeur, Eric. "The Forgotten Tool: The Design and Use of Molecular Models." *Social Studies of Science* 27 (1997): 7–40.

Frehner, Brian. *Finding Oil: The Nature of Petroleum Geology.* Lincoln: University of Nebraska Press, 2011.

Gait, George B. *Microfilming Maps of Abandoned Anthracite Mines: Mines in the Lackawanna Basin, Northern Anthracite Field.* Bureau of Mines Information Circular 8453. [Washington, DC]: US Department of the Interior, Bureau of Mines, 1970.

———. *Microfilming Maps of Abandoned Anthracite Mines: Mines in the Southern Anthracite Field.* Bureau of Mines Information Circular 8779. [Washington, DC]: US Department of the Interior, Bureau of Mines, 1978.

———. *Microfilming Maps of Abandoned Anthracite Mines: Mines in the Western Middle Anthracite Field.* Bureau of Mines Information Circular 8519. [Washington, DC]: US Department of the Interior, Bureau of Mines, 1971.

Gilbert, Chester G. "Coal Products: An Object Lesson in Resource Administration." *United States National Museum Bulletin* 102, no. 1 (1917): 1–16.

———. "Observations on the Use of Models in the Educational Work of Museums." *Museum Work* 2, no. 3 (December 1919): 90–92.

Glasscock, C. B. *War of the Copper Kings: Builders of Butte and Wolves of Wall Street.* New York: Bobbs-Merrill, 1935.

Gow, Paul A. "Mine Surveying Methods Employed at Butte, Montana." *Engineering and Mining Journal* 90 (1910): 1209–1211.

Graeme, Richard W. "Bisbee, Arizona's Dowager Queen of Mining Camps: A Look at Her First 50 Years." In *History of Mining in Arizona,* edited by J. Michael Canty and Michael N. Greeley, 51–75. Vol. 1. Tucson: Mining Club of the Southwest Foundation, 1987.

"Graphic Indication of Drilling." *Engineering and Mining Journal* 89 (February 1910): 453.

Graton, L. C. "Seventy-five Years of Progress in Mining Geology." In *Seventy-five Years of Progress in the Mineral Industry, 1871–1946,* edited by A. B. Parsons, 1–39. New York: American Institute of Mining / Metallurgical Engineers, 1947.

Greene, Mott T. *Geology in the Nineteenth Century: Changing Views of a Changing World.* Ithaca, NY: Cornell University Press, 1982.

Gutfeld, Arnon. "The Murder of Frank Little: Radical Labor Agitation in Butte, Montana, 1917." *Labor History* 10, no. 2 (1969): 177–192.

———. "The Speculator Disaster in 1917: Labor Resurgence at Butte, Montana." *Arizona and the West* 11, no. 1 (1969): 27–38.

Hall, Benjamin James. *Blue Printing and Modern Plan Copying for the Engineer and Architect, the Draughtsman and the Print Room Operative.* London: Sir I. Pitman & Sons, 1921.

Harden, John H., and Edward B. Harden. "The Construction of Maps in Relief." *Transactions of the American Institute of Mining Engineers* 16 (1888): 279–301.

Harden, O. B. "Topographical and Geological Modelling." *Transactions of the American Institute of Mining Engineers* 10 (1882): 264–267.

Harley, J. Brian. *The New Nature of Maps: Essays in the History of Cartography.* Edited by Paul Laxton. Baltimore: Johns Hopkins University Press, 2001.

Harrison, T. S., and H. C. Zulch. "Court Maps and Models." *Mines and Minerals* 29 (September 1908): 49–54.

Hayes, J. L. "Anaconda Copper Mining Co.'s Mapping Practice." *Engineering and Mining Journal-Press* 116 (1923): 841–843.

"Head-Frame Models Made of Paper." *Mines and Minerals* 30 (February 1910): 401–402.

Heller, Dean, ed. *Political History of Nevada, 1996.* 10th ed. Carson City, NV: State Printing Office, 1997.

Henderson, Kathryn. *On Line and On Paper: Visual Representations, Visual Culture, and Computer Graphics in Design Engineering.* Cambridge, MA: MIT Press, 1999.

Henson, Pamela M. "'Objects of Curious Research': The History of Science and Technology at the Smithsonian." *Isis* 90, no. Supplement (1999): S249–S269.

Hindle, Brooke. *Emulation and Invention*. New York: New York University Press, 1981.

Hittel, John S. *Hittel on Gold Mines and Mining*. Quebec [Quebec City?]: G. & G. E. Desbarats, 1864.

Hoffman, John N. *Anthracite in the Lehigh Region of Pennsylvania, 1820–45*. Contributions from the Museum of History and Technology 72. Washington, DC: Smithsonian Institution Press, 1968.

Hoover, Herbert. *The Memoirs of Herbert Hoover: Years of Adventure, 1874–1920*. New York: Macmillan, 1951.

Hopkins, C. V. "United Verde Mine Model." *Mines and Minerals* 30 (March 1910): 501–502.

House Select Committee on Existing Labor Troubles in Pennsylvania. *Labor Troubles in the Anthracite Regions of Pennsylvania, 1887–1888*. 50th Congress, 2nd Session, House of Representatives Report 4147. Washington, DC: GPO, 1889.

Hovis, Logan, and Jeremy Mouat. "Miners, Engineers, and the Transformation of Work in the Western Mining Industry, 1880–1930." *Technology and Culture* 37, no. 3 (July 1996): 429–456.

Hudson Coal Company. *The Story of Anthracite*. New York: Hudson Coal Company, 1932.

Hughes, Thomas P. "The Evolution of Large Technological Systems." In *The Social Construction of Technological Systems: New Directions in the Sociology and History of Technology*, edited by Wiebe E. Bijker, Thomas P. Hughes, and Trevor Pinch, 51–82. Cambridge, MA: MIT Press.

———. *Networks of Power: Electrification in Western Society, 1880–1930*. Baltimore: Johns Hopkins University Press, 1983.

"Importance of Accurate Mine Surveys." *The Colliery Engineer* 8 (May 1888): 228.

"An Interesting Mine Model." *Mining and Scientific Press* 29 (August 1874): 136.

Jackson, Carlton. *The Dreadful Month*. Bowling Green, OH: Bowling Green University Popular Press, 1982.

Jensen, Vernon H. *Heritage of Conflict: Labor Relations in the Nonferrous Metals Industry up to 1930*. Ithaca, NY: Cornell University Press, 1950.

Johnson, J. B. *The Theory and Practice of Surveying*. 16th ed. New York: John Wiley & Sons, 1908.

Jones, William. *The Treasures of the Earth; or, Mines, Minerals, and Metals*. London: Frederick Warne and Co., 1868.

Jordanova, Ludmilla. "Material Models as Visual Culture." In *Models: The Third Dimension of Science*, edited by Soraya de Chadarevian and Nick Hopwood, 443–451. Stanford, CA: Stanford University Press, 2004.

Keiner, Christine. "Modeling Neptune's Garden: The Chesapeake Bay Hydraulic Model, 1965–1984." In *The Machine in Neptune's Garden: Historical Studies on Technology and the Marine Environment*, edited by Helen M. Rozwadowski and

David K. van Keuren, 273–314. Sagamore Beach, MA: Science History Publications, 2004.

Kilander, Ginny. "Transits, Timbers, & Tunnels: The Legacy of Colorado Inventor David W. Brunton." In *Enterprise and Innovation in the Pikes Peak Region,* edited by Tim Blevins, 65–101. Colorado Springs, CO: Pikes Peak Library District with Dream City Vision 2020, 2011.

Knight, Nelson. "This Old House: Dickson-Gardner-Wolfe Mansion." *The Capitol Hill Neighborhood Council Bulletin* [Salt Lake City, UT], no. 49 (May 2005): 1–2.

Kurie, F. M. "Glass Models of Portland Mine: A Convenient Means for Plainly Illustrating Mine-Workings, Geological Features, Etc." *Mines and Minerals* 24 (February 1904): 307.

Lacy, John C. "Pursuing Segments of the Glass Snake: The Battle for the Big Jim." In *History of Mining in Arizona,* edited by J. Michael Canty and Michael N. Greeley, 263–285. Vol. 2. Tucson, AZ: Mining Club of the Southwest Foundation / The American Institute of Mining Engineers Tucson Section, 1991.

Lake, Mack C. "Mine Models Made of Celluloid Sheets." *Engineering and Mining Journal* 99 (April 1915): 737–738, 957.

Lane, H. M. "Plumbing Deep Shafts of the Tamarack Mine." *Mines and Minerals* 22 (January 1902): 247–248.

Lankton, Larry. *Cradle to Grave: Life, Work, and Death at the Lake Superior Copper Mines.* New York: Oxford University Press, 1991.

———. *Hollowed Ground: Copper Mining and Community Building on Lake Superior, 1840s–1990s.* Detroit, MI: Wayne State University Press, 2010.

Lankton, Larry D., and Charles K. Hyde. *Old Reliable: An Illustrated History of the Quincy Mining Company.* Hancock, MI: Quincy Mine Hoist Association, 1982.

Lawson, Andrew C., and Harry Fielding Reid. *The California Earthquake of April 18, 1906: Report of the State Earthquake Investigation Commission.* Washington, DC: Carnegie Institution of Washington, 1908.

Layton, Edwin T. *The Revolt of the Engineers: Social Responsibility and the American Engineering Profession.* Cleveland, OH: Press of Case Western Reserve University, 1971.

LeCain, Timothy J. *Mass Destruction: The Men and Giant Mines That Wired America and Scarred the Planet.* New Brunswick, NJ: Rutgers University Press, 2009.

Lehman, A. E. "Topographical Models: Their Construction and Uses." *Transactions of the American Institute of Mining Engineers* 14 (1886): 439–455.

Leifchild, John R. *Cornwall: Its Mines and Miners; with Sketches of Scenery; Designed as a Popular Introduction to Metallic Mines.* London: Longman Brown Green and Longmans, 1855.

Lesley, J. P. "Biographical Notice of Charles A. Ashburner." *Transactions of the American Institute of Mining Engineers* 18 (1890): 365–370.

———. *Manual of Coal and its Topography.* Philadelphia: J. B. Lippincott, 1856.

———. "Topographical Models or Relief-Maps." *Science* 7, no. 154 (1886): 58.

Lietze, Ernst. *Modern Heliographic Processes: A Manual of Instruction in the Art of*

Reproducing Drawings, Engravings, Manuscripts, Etc., by the Action of Light; for the Use of Engineers, Architects, Draughtsmen, Artists, and Scientists. New York: D. Van Nostrand, 1888.

Lindley, Curtis H. "A Problem in Extralateral Rights on Secondary Veins." *California Law Review* 1 (1913): 427–438.

———. *A Treatise on the American Law Relating to Mines and Mineral Lands Within the Public Land States and Territories and Governing the Acquisition and Enjoyment of Mining Rights in Lands of the Public Domain (aka Lindley on Mines).* 3rd ed. San Francisco: Bancroft-Whitney, 1914.

Linforth, F[rank] A. "Application of Geology to Mining in the Ore Deposits at Butte Montana." In *Ore Deposits of the Western States: The Lindgren Volume,* edited by John Wellington Finch and the Committee on the Lindgren Volume, 695–701. New York: The American Institute of Mining and Metallurgical Engineers, 1933.

———. "Applied Geology in the Butte Mines." *Transactions of the American Institute of Mining Engineers* 46 (1914): 110–122.

Linforth, F[rank] A., and E. B. Milburn. "Geology Applied to Mining." *Engineering and Mining Journal* 91 (April 1911): 664–667.

Lingenfelter, Richard E. *The Hardrock Miners: A History of the Mining Labor Movement in the American West, 1863–1893.* Berkeley: University of California Press, 1974.

Linklater, Andro. *Measuring America: How the United States Was Shaped by the Greatest Land Sale in History.* New York: Plume Books / Penguin, 2002.

Locke, Augustus. "The Geology of the Tonopah Mining District." *Transactions of the American Institute of Mining Engineers* 43 (1912): 157–166.

Lubar, Steven. "Representation and Power." *Technology and Culture* 36, no. 2, Special Issue (1995): S54–S81.

Lubar, Steven, and Kathleen M. Kendrick. *Legacies: Collecting America's History at the Smithsonian.* Washington, DC: Smithsonian Institution Press, 2001.

Lyman, Benjamin Smith. "Folds and Faults in Pennsylvania Anthracite-Beds." *Transactions of the American Institute of Mining Engineers* 25 (1896): 327–369.

———. "On the Importance of Surveying in Geology." *Transactions of the American Institute of Mining Engineers* 1 (1873): 183–192.

Malone, Michael P. *The Battle for Butte: Mining and Politics on the Northern Frontier, 1864–1906.* Seattle: University of Washington Press, 1981.

Marcosson, Isaac F. *Anaconda.* New York: Dodd, Mead, 1957.

McCracken, Robert D. *A History of Tonopah, Nevada.* Tonopah, NV: Nye County Press, 1990.

McDonald, Philip B. "Eckley Brinton Coxe." *Dictionary of American Biography* Base Set. Reproduced in *Biography Resource Center.* Farmington Hills, MI: Thomson Gale, 2007.

McGivern, James G. "Polytechnic College of Pennsylvania: A Forgotten College." *Journal of Engineering Education* 52, no. 2 (November 1961): 106–112.

McNair, James B. *With Rod and Transit: The Engineering Career of Thomas S. McNair (1824–1901)*. Los Angeles: Published by the Author, 1951.

Meier, John W. "Mines on the Lahn in Nassau, Germany." *Engineering and Mining Journal* 54 (1892): 414–415.

Meyer, Charles, Edward P. Shea, Charles C. Goddard Jr., and Staff. "Ore Deposits at Butte, Montana." In *Ore Deposits of the United States, 1933–1967: The Graton-Sales Volume*, 1st ed., vol. 2, edited by John D. Ridge, 1373–1416. New York: The American Institute of Mining, Metallurgical, and Petroleum Engineers, 1968.

Miller, Charles W. Jr. *Stake Your Claim! The Tale of America's Enduring Mining Laws*. Tucson, AZ: Westernlore Press, 1991.

Miller, Donald L., and Richard E. Sharpless. *The Kingdom of Coal: Work, Enterprise, and Ethnic Communities in the Mine Fields*. Philadelphia: University of Pennsylvania Press, 1985.

Miller, W. St. J. "A Satisfactory Mine Model." *Mining and Scientific Press* 103 (1911): 779.

Milton, Maxwell C. "A Mine Model." *Mining and Scientific Press* 105 (1912): 338.

"Mine Model." *Engineering and Mining Journal* 75 (1903): 904.

"Mine Model." *Mining and Scientific Press* 34 (June 1877): 337.

"Mine Model for Inclined Veins." *Engineering and Mining Journal* 90 (December 1910): 1243.

"Mine Model of Vertical Glass Plates." *Engineering and Mining Journal* 99 (January 1915): 236–237.

"Mine Model or Exhibit." *Engineering and Mining Journal* 75 (1903): 757.

"Mine Models in Tonopah Apex Litigation." *Engineering and Mining Journal* 99 (1915): 850.

Molella, Arthur P. "The Museum That Might Have Been: The Smithsonian's National Museum of Engineering and Industry." *Technology and Culture* 32, no. 2, pt. 1 (April 1991): 237–263.

Morgan, Mary S., and Margaret Morrison, eds. *Models as Mediators: Perspectives on Natural and Social Science*. Cambridge, UK: Cambridge University Press, 1999.

"MSHA Announces Grants to Digitize Abandoned Mine Maps." *Pit & Quarry* 96 (February 2004): 55.

"MSHA Finds New Map of Flooded Mine." *Coal Age* 107 (October 2002): 6.

Munroe, H. S. "A Summer School of Practical Mining." *Transactions of the American Institute of Mining Engineers* 9 (1881): 664–671.

Murphy, John N. "Update on the Continuing Functions of the Former US Bureau of Mines." *Mining Engineering* 49 (June 1997): 87–89.

Murphy, Mary. *Mining Cultures: Men, Women, and Leisure in Butte, 1914–41*. Urbana: University of Illinois Press, 1997.

Neff, John L. "The Law of the Apex: A Continuing Enigma." *Rocky Mountain Mineral Law Institute* 18 (1973): 387–414.

"New Use for a Mine Model." *Mining and Scientific Press* 87 (October 1903): 247.

Nolan, Thomas B. "The Underground Geology of the Tonopah Mining District, Nevada." *University of Nevada Bulletin* 29, no. 5 (1935): 1–49.

North, Edmund D. "Glass Mine-Models." *Transactions of the American Institute of Mining Engineers* 40 (1910): 755–759.

Nystrom, Eric. "'Without Doubt the Most Accurate': Underground Surveying and the Development of Mining Engineering in the Pennsylvania Anthracite Region." *Pennsylvania Legacies* 9, no. 2 (November 2009): 20–25.

O'Brien, J. P., ed. *History of the Bench and Bar of Nevada*. San Francisco: Bench and BarPublishing Company, 1913.

Ochs, Kathleen H. "The Rise of American Mining Engineers: A Case Study of the Colorado School of Mines." *Technology and Culture* 33, no. 2 (April 1992): 278–301.

Office of Surface Mining Reclamation and Enforcement. *Mine Map Repositories: A Source of Mine Map Data*. US Department of the Interior, Office of Surface Mining Reclamation and Enforcement, Program Information Development, April 1989.

Owens, W. L. "Surveying and Mapping." *Coal Age* 12 (1917): 960–961.

Parry, William T. *All Veins, Lodes, and Ledges Throughout Their Entire Depth: Geology and the Apex Law in Utah Mines*. Salt Lake City: University of Utah Press, 2004.

Parsons, A. B. "History of the Institute." In *Seventy-five Years of Progress in the Mineral Industry, 1871–1946*, edited by A. B. Parsons, 403–529. New York: American Institute of Mining and Metallurgical Engineers, 1947.

———. *The Porphyry Coppers*. New York: American Institute of Mining and Metallurgical Engineers, 1933.

Paul, Rodman Wilson. "Colorado as a Pioneer of Science in the Mining West." *The Mississippi Valley Historical Review* 47, no. 1 (June 1960): 34–50.

Paul, Rodman W[ilson], and Elliott West. *Mining Frontiers of the Far West, 1848–1880*. 2nd ed. Albuquerque: University of New Mexico Press, 2001.

Peele, Robert, ed. *Mining Engineers' Handbook*. 1st ed. New York: John Wiley and Sons, 1918.

Peele, Robert, and John A. Church, eds. *Mining Engineers' Handbook*. 3rd ed. New York: John Wiley & Sons, 1941.

"The Pennsylvania M. Co. vs. Grass Valley Exploration Co. Litigation." *Mining and Scientific Press* 83 (August 1901): 41, 44.

Penrose, R. A. F. "Notes on the State Exhibits in the Mines and Mining Building at the World's Columbian Exposition, Chicago." *Journal of Geology* 1, no. 5 (1893): 457–470.

Perry, Vincent D., and Charles Meyer. "Reno H. Sales." In *Ore Deposits of the United States, 1933–1967: The Graton-Sales Volume*, 1st ed., vol. 1, edited by John D. Ridge, xxi–xxiii. New York: The American Institute of Mining, Metallurgical, and Petroleum Engineers, 1968.

Pogue, Joseph E. "Sulphur: An Example of Industrial Independence." *United States National Museum Bulletin* 102, no. 3 (1917): 1–10.

"Proceedings of the Lake Superior Meeting." *Transactions of the American Institute of Mining Engineers* 27 (1898): xxx–xlix.

Quecreek Miners. *Our Story: 77 Hours that Tested Our Friendship and Our Faith.* As told to Jeff Goodell. New York: Hyperion, 2002.

Ramsey, Robert H. *Men and Mines of Newmont: A Fifty-Year History.* New York: Octagon Books, 1973.

Raymond, Rossiter W. "Biographical Notice of Eckley B. Coxe." *Transactions of the American Institute of Mining Engineers* 25 (1896): 446–476.

———. "The Law of the Apex." *Transactions of the American Institute of Mining Engineers* 12 (1884): 387–444.

———. "The Law of the Apex: Appendix." *Transactions of the American Institute of Mining Engineers* 12 (1884): 677–688.

———. "Review of 'A Treatise on the American Law Relating to Mines and Mineral Lands, Etc.'" *Engineering and Mining Journal* 97 (April 1914): 817.Read, Thomas Thornton. *The Development of Mineral Industry Education in the United States.* New York: American Institute of Mining and Metallurgical Engineers, 1941.

Reist, H. G. "Blue-Printing by Electric Light." *Transactions of the American Society of Mechanical Engineers* 22 (1901): 888–898.

Report on the Progress and Condition of the United States National Museum for the Year Ending June 30, 1910. Washington, DC: GPO, 1911.

Report on the Progress and Condition of the United States National Museum for the Year Ending June 30, 1911. Washington, DC: GPO, 1912.

Report on the Progress and Condition of the United States National Museum for the Year Ending June 30, 1912. Washington, DC: GPO, 1913.

Report on the Progress and Condition of the United States National Museum for the Year Ending June 30, 1913. Washington, DC: GPO, 1914.

Report on the Progress and Condition of the United States National Museum for the Year Ending June 30, 1914. Washington, DC: GPO, 1915.

Report on the Progress and Condition of the United States National Museum for the Year Ending June 30, 1919. Washington, DC: GPO, 1920.

Reports of the Inspectors of Coal Mines of the Anthracite Coal Regions of Pennsylvania for the Year 1870. Harrisburg, PA: B. Singerly, State Printer, 1871.

Reports of the Inspectors of Coal Mines of the Anthracite Coal Regions of Pennsylvania for the Year 1871. Harrisburg, PA: B. Singerly, State Printer, 1872.

Reports of the Inspectors of Mines of the Anthracite Coal Regions of Pennsylvania for the Year 1872. Harrisburg, PA: Benjamin Singerly, State Printer, 1873.

Reports of the Inspectors of Mines of the Anthracite Coal Regions of Pennsylvania for the Year 1873. Harrisburg, PA: Benjamin Singerly, State Printer, 1874.

Reports of the Inspectors of Mines of the Anthracite Coal Regions of Pennsylvania for the Year 1874. Harrisburg, PA: B. F. Meyers, State Printer, 1875.

Reports of the Inspectors of Mines of the Anthracite Coal Regions of Pennsylvania for the Year 1875. Harrisburg, PA: B. F. Meyers, State Printer, 1876.

Reports of the Inspectors of Mines of the Anthracite Coal Regions of Pennsylvania for the Year 1876. Harrisburg, PA: B. F. Meyers, State Printer, 1877.

Reports of the Inspectors of Mines of the Anthracite Coal Regions of Pennsylvania for the Year 1877. Harrisburg, PA: Lane S. Hart, State Printer, 1878.

Reports of the Inspectors of Mines of the Anthracite Coal Regions of Pennsylvania for the Year 1878. Harrisburg, PA: Lane S. Hart, State Printer, 1879.

Richards, F. B. "Mining." In *Engineering as a Career: A Series of Papers by Eminent Engineers,* edited by F. H. Newell and C. E. Drayer, 149–158. New York: D. Van Nostrand, 1916.

Richards, R. H. "American Mining Schools." *Transactions of the American Institute of Mining Engineers* 15 (1887): 309–340.

———. "American Mining Schools." *Transactions of the American Institute of Mining Engineers* 15 (1887): 809–819.

Rickard, T. A. "Geology Applied to Mining." *Mining and Scientific Press* 100 (1910): 479–481.

———. *A History of American Mining.* New York: McGraw-Hill, 1932.

———. *Interviews with Mining Engineers.* San Francisco: Mining and Scientific Press, 1922.

Righter, Robert W. *The Battle over Hetch Hetchy: America's Most Controversial Dam and the Birth of Modern Environmentalism.* New York: Oxford University Press, 2005.

Robinson, B. W. "Mine Maps: A Paper Showing What Is Required on a Good Map, and Why Complete Maps Are a Necessity." *The Colliery Engineer* 9 (April 1889): 198.

Rockwell, John A. *A Compilation of Spanish and Mexican Law, in Relation to Mines, and Titles to Real Estate, in Force in California, Texas, and New Mexico.* Vol. 1. New York: John S. Voorhies, 1851.

Rothwell, Richard P. "The Mechanical Preparation of Anthracite." *Transactions of the American Institute of Mining Engineers* 3 (1875): 134–144.

Rouse, Hunter. "Weisbach, Julius Ludwig." In *Dictionary of Scientific Biography,* edited by Charles Coulston Gillispie, 232. Vol. 14. New York: Charles Scribner's Sons, 1976.

Rydell, Robert W. *All the World's A Fair: Visions of Empire At American Industrial Expositions, 1876–1916.* Chicago: University of Chicago Press, 1984.

Sales, Reno H. "Ore Discovery and Development of Fundamental Importance: How Anaconda's Large Geological Staff Is Employed." *Engineering and Mining Journal* 128 (1929): 277–279.

———. *Underground Warfare at Butte.* Caldwell, ID: Caxton Printers, 1964.

Sarkar, Dibya. "States to Make Digital Maps for Mines." *Federal Computer Week* 18 (January 2004): 14.

Schneider, Jimmie. *Quicksilver: The Complete History of Santa Clara County's New Almaden Mine.* San Jose, CA: Zella Schneider, 1992.

Scott, Dunbar D., and others. *The Evolution of Mine-Surveying Instruments*. New York: American Institute of Mining Engineers, 1902.

Searls, Fred Jr. "The Consultant." In *Andrew C. Lawson: Scientist, Teacher, Philosopher*, edited by Francis E. Vaughan, 205–211. Glendale, CA: A. H. Clark, 1970.

Severy, C. L. "Making Mine Models." *Mining and Scientific Press* 104 (1912): 381.

Shamberger, Hugh A. *The Story of Candelaria and its Neighbors: Columbus, Metallic City, Belleville, Marietta, Sodaville and Coaldale, Esmeralda and Mineral Counties, Nevada*. Carson City, NV: Nevada Historical Press, 1978.

Shamel, Charles H. *Mining, Mineral, and Geological Law: A Treatise on the Law of the United States*. New York: Hill Publishing Company, 1907.

Shea, Patrick. "Against All Odds: Coxe Brothers & Company, Inc. and the Struggle to Remain Independent." Paper presented at Mining History Association Conference, Scranton, PA, June 2005.

Sherwood, Don H. "The Extralateral Right: A Last Hurrah?" *Rocky Mountain Mineral Law Institute* 35 (1989): 12.1–12.34.

Shor, Elizabeth Noble. "Winchell, Horace Vaughn." In *American National Biography Online*. New York: Oxford University Press, 2000.

Smith, Duane A. *Mining America: The Industry and the Environment, 1800–1980*. Lawrence: University Press of Kansas, 1987; reprint, Niwot: University Press of Colorado, 1993.

Smith, Duane A., Karen A. Vendl, and Mark A. Vendl. *Colorado Goes to the Fair: World's Columbian Exposition, Chicago, 1893*. Albuquerque: University of New Mexico Press, 2011.

Smith, Grant H., and Joseph V. Tingley. *The History of the Comstock Lode*. Reno: Nevada Bureau of Mines and Geology, in association with University of Nevada Press, 1998.

Sopwith, T. *A Treatise on Isometrical Drawing, as Applicable to Geological and Mining Plans, Picturesque Delineations of Ornamental Grounds, Perspective Views and Working Plans of Buildings and Machinery, and to General Purposes of Civil Engineering; With Details of Improved Methods of Preserving Plans and Records of Subterranean Operations in Mining Districts*. London: John Weald, Taylor's Architectural Library, 1834.

———. *Description of a Series of Elementary Geological Models*. London: R. J. Mitchell & Sons, 1875.

Sopwith, Thomas. *Description of a Series of Geological Models, Illustrating the Nature of Stratification, Valleys of Denudation, Succession of Coal Seams in the Newcastle Coal Field, the Effects Produced by Faults or Dislocation of the Strata, Intersection of Mineral Veins, &c.* Newcastle-upon-Tyne: Printed for the Author, by J. Blackwell and Co., 1841.

Spence, Clark C. *Mining Engineers and the American West: The Lace-Boot Brigade, 1849–1933*. 1970; reprint, Moscow: University of Idaho Press, 1993.

Spurr, J. E. "Geology and Ore Deposits at Tonopah, Nev." *Economic Geology* 10 (1915): 751–760.

———. *Geology of the Tonopah Mining District, Nevada.* US Geological Survey Professional Paper 42. Washington, DC: GPO, 1905.

St. Clair, David J. "New Almaden and California Quicksilver in the Pacific Rim Economy." *California History* 73 (1994): 278–295.

Stoek, H. H. "Mine Models." *Transactions of the American Institute of Mining Engineers* 58 (1918): 25–35.

Storms, William H. *Timbering and Mining: A Treatise on Practical American Methods.* 1st ed. New York: McGraw-Hill, 1909.

Street, Reuben. "Mine Surveying." *The Colliery Engineer* 8 (March 1888): 177.

Suggs, George. *Colorado's War on Militant Unionism.* Norman: University of Oklahoma Press, 1972.

"A Supplement to an Act, entitled 'An Act providing for the health and safety of persons employed in coal mines' . . . " In Laws of the General Assembly of the State of Pennsylvania Passed at the Session of 1876. (Harrisburg, PA: B. F. Meyers, State Printer, 1876), 130.

Third Biennial Report of the State Mine Inspectors, to the Governor of the State of Iowa, for the Years 1886 and 1887. Des Moines, IA: Geo. E. Roberts, State Printer, 1888.

Transcript of Record, Supreme Court of the United States, Nos. 636 (116) and 637 (117). The Richmond Mining Company of Nevada, et al., Appellants, vs. The Eureka Consolidated Mining Company, Appeals from the Circuit Court of the United States of the District of Nevada, filed January 17, 1878. Records and Briefs of the Supreme Court.

Uhrin, David C. *Correlating Microfilm Mine Maps with Topographic Maps.* Bureau of Mines Information Circular 8762. [Washington, DC]: US Department of the Interior, Bureau of Mines, 1977.

"Unit Blocks for Mine Model." *Engineering and Mining Journal* 97 (May 1914): 1049–1050.

"The Utah Consolidated Mining Company." *Engineering and Mining Journal* 82 (September 1906): 488–489.

Vincenti, Walter G. *What Engineers Know and How They Know It: Analytical Studies from Aeronautical History.* Baltimore: Johns Hopkins University Press, 1990.

von Bitter, Peter H. "The Brunton Pocket Transit, A One Hundred Year Old North American Invention." *Earth Sciences History* 14, no. 1 (1995): 98–102.

Walker, William S. *A Living Exhibition: The Smithsonian and the Transformation of the Universal Museum.* Amherst: University of Massachusetts Press, 2013.

Wallace, Anthony F. C. *St. Clair: A Nineteenth-Century Coal Town's Experience with a Disaster-Prone Industry.* Ithaca, NY: Cornell University Press, 1988.

Weed, Walter Harvey. *Geology and Ore Deposits of the Butte District, Montana.* US Geological Survey Professional Paper 74. Washington: GPO, 1912.

Weisbach, Julius. *Neue Markscheidekunst und ihre Anwedung auf Bergmännische Anlagen.* Braunschweig: Verlag von Friedrich Vieweg und Sohn, 1859.

———. *Theoretical Mechanics: With an Introduction to the Calculus: Designed as a*

Text-Book for Technical Schools and Colleges, and for the Use of Engineers, Architects, Etc.; Translated from the Fourth Augmented and Improved German Edition. Translated by Eckley B. Coxe. New York: D. Van Nostrand, 1875.

Wetherill, J. Price. "An Outline of Anthracite Coal Mining in Schuylkill County, Pa." *Transactions of the American Institute of Mining Engineers* 5 (1877): 402–422.

Whaite, Ralph H. *Microfilming Maps of Abandoned Anthracite Mines: Mines of the Eastern Middle Field.* Bureau of Mines Information Circular 8274. [Washington, DC]: United States Department of the Interior, Bureau of Mines, 1965.

Wilkins, Thurman. *Clarence King: A Biography.* Rev. Ed. Albuquerque: University of New Mexico Press, 1988.

Williams, Rosalind. *Notes on the Underground: An Essay on Technology, Society, and the Imagination.* Cambridge, MA: MIT Press, 1990.

Winchell, Horace V. "Review of 'Lindley on Mines.'" *Economic Geology* 9, no. 6 (September 1914): 598–602.

Winchell, N. H., and H. V. Winchell. *The Iron Ores of Minnesota: Their Geology, Discovery, Development, Qualities, and Origin, and Comparison With Those of Other Iron Districts.* Geological and Natural History Survey of Minnesota Bulletin 6. Minneapolis: Harrison and Smith, State Printers, 1891.

Winchester, Simon. *The Map That Changed The World: William Smith and the Birth of Modern Geology.* New York: HarperCollins, 2001.

Wolensky, Robert P., and Joseph M. Keating. *Tragedy at Avondale: The Causes, Consequences, and Legacy of the Pennsylvania Anthracite Coal Industry's Most Deadly Mining Disaster, September 6, 1869.* Easton, PA: Canal History and Technology Press, 2008.

Wolensky, Robert P., Kenneth C. Wolensky, and Nicole H. Wolensky. *The Knox Mine Disaster, January 22, 1959: The Final Years of the Northern Anthracite Industry and the Effort to Rebuild a Regional Economy.* Harrisburg, PA: Pennsylvania Historial and Museum Commission, 1999.

———. *Voices of the Knox Mine Disaster: Stories, Remembrances, and Reflections of the Anthracite Coal Industry's Last Major Catastrophe, January 22, 1959.* Harrisburg, PA: Pennsylvania Historical and Museum Commission, 2005.

Wood, Denis. *The Power of Maps.* New York: Guilford Press, 1992.

"Wood and Cloth Mine Model." *Engineering and Mining Journal* 95 (February 1913): 419–420.

Wyman, Mark. *Hard Rock Epic: Western Miners and the Industrial Revolution, 1860–1910.* Berkeley: University of California Press, 1979.

Yaste, George L. "Care of Mine Maps." *Coal Age* 10 (1916): 223.

Yates, JoAnne. *Control through Communication: The Rise of System in American Management.* Baltimore: Johns Hopkins University Press, 1989.

Zulch, W. G. "Stope Model." *Mines and Minerals* 32 (1912): 491.

INDEX

Page numbers in italics refer to illustrations.

Carbondale mine disaster, 64–65
card catalog systems, 35–36
Carpenter, Jay, 162
Centennial Exhibition of 1876 (Philadelphia), 201
center-pins, 24
Centralia fire, 223
Chamberlin, John R., 132–33
Chandler, John W., 159, 161–62, 169
Chicago World's Fair of 1893, 202–3
cinnabar. See mercury mining
clinometers, 20, 96
coal mining: in American history, 5, 7; development of underground mine maps in the industrialization of mining, 40–41, 44–46; human and environmental costs of, 13. See also anthracite coal mining
coal workers: Eckley Coxe's treatment of, 69–70
Colby, William E., 157, 189, 262n117
Cole, Walter D., 159
"Colorado Gold Mine" model, 203
Columbia College School of Mines, 47, 48, 199–200, 206
Columbian Exposition of 1893 (Chicago), 202–3
commercial mine models: donated to schools, 192–93, 195
compasses, 20–21, 22, 37, 39
composite mine maps: description of, 27–28; in the development of the Anaconda System, 91, 92; used in the Jim Butler v. West End case, 170
Comstock Lode, 128
Comstock Lode maps, 29
Consolidated Virginia Mine, 128
Consolidation Coal Company, 208–9
copper mining: in American history, 5, 7; in Butte, 86; impact on the development of the United States, 13; negative models and, 140–43. See also Anaconda Copper Mining Company
Copper Queen Mine, 134, 147
copper tailings: negative modeling of, 142–43
Coryell, Martin, 59
Coxe, Eckley B.: American Institute of Mining Engineering and, 48, 59, 66;

biographical overview, 54; death of, 77; evolution of the Drifton Colliery maps through the 1870s, 70–76; on the "ideal" mining engineer, 48; impact on mining engineering, mine mapping, and the visual culture of mining, 54; innovations in mining practice, 69–70; innovations in underground surveying, 25, 66–69; management of Coxe Brothers & Co., 65–66, 69–70; mining education of, 66, 67; political and civic career, 66–67; professional career, 66; professionalization of mining engineering and, 65–66; railroads and, 70; treatment of coal workers, 69–70. See also Drifton Colliery
Coxe, Tench, 54, 239n35
Coxe Brothers & Company: Eckley Coxe's death and, 77; Eckley Coxe's management of, 65–66, 69–70; further developments in mapping the early twentieth century, 77, 79–83; history of, 239n35; map filing system, 36. See also Drifton Colliery
Crane, Walter R., 194, 195–97, 198
Cripple Creek block model, 124, 125
Cross Creek Colliery, 239n35
crosscuts: on Anaconda System maps, 97
culm, 57

Daniels, Joseph, 200
Day, Marcus, 86, 88, 89, 90–91
Deister Machine Company, 214
Delaware, Susquehanna, and Schuylkill Railroad, 70
Della S. Consolidated Mining Company, 140
deputy mineral surveyors, 18
De Re Metallica (Agricola), 37, 38, 66
Dewey, Frederick P., 130, 201–2
Dickson, William H.: biographical overview, 157–58; in the Jim Butler v. West End case, 157, 158, 166, 167–69, 171–74, 185, 189; later years and death of, 256n30; Silver King v. Conkling case, 188
digitization: of mining maps, 226
dips, 164, 254n12

Street, Reuben, 49
strikes: defined, 254n12; *Jim Butler v. West End* case and, 164, 178
striking workers. *See* labor strikes
study models, 115–16
subsidence, 223
"sun print" map copies, 30
surface property and claim maps, 18
surface surveying, 18, 19–20
surveyor's chains, 25, 67
surveyor's notebooks, 25, 96–97
surveyor's stations: in Eckley Coxe's 1870s maps of the Drifton Colliery, 71, 72, 73, 74, 75, 76; in underground surveying, 22–25
surveyor's tape measure, 25, 67, 69
synclines, 56, 163

technical mine models: aesthetics of, 148; block models, 117–25; categories of, 117; combination models, 143; defined, 114; distinguished from display models, 114; functioning of, 114–15; in the *Jim Butler v. West End* case, 149–50, 165–74, 175, 177–78, 180–81, 185, 188, 190–91; in the late nineteenth and early twentieth centuries, 116–17; negative models, 117, 135–43; power of, 113; purpose of, 116; representational work and technical communication, 190–91; sectional models, 117, 125–35; uses of, 113, 143–48; the visual culture of mining and, 147–48. *See also individual model types*
terminal edges, 182–83, 184, 185
theodolites: description of, 21–22; use in conducting an underground survey, 24, 25
theoretical models, 114
Tonopah Belmont Mine, 151, 158
Tonopah Extension Mining Company, 134, 162
Tonopah Historic Mining Park, 253n1
Tonopah Mining Company, 145, 151, 158, 189
Tonopah (NV): history of mining in, 150–52. See also *Jim Butler v. West End* case

Tonopah 76 Mining Company, 163
topographic block models, 118–21, 193
Torch Lake copper tailings, 142–43
training mines, 193
transits: description of, 21–22, 23; use in conducting an underground survey, 24, 25
tube storage systems, 33

underground elevation: incorporation in block models, 121–22
underground mine maps/mapping: Anaconda System, 85–86 (*see also* Anaconda System); appearance of technical literature on, 49; blueprinting and blueprints, 29–33, 233n34; development in the industrialization of mining in America, 40–46, 83–84; drafting of maps, 25–29; elevation measurements, 75, 76; evolution of Eckley Coxe's maps through the 1870s, 70–76; filing and organizational systems, 35–37; impact of Eckley Coxe on, 54; impact of Pennsylvania mining regulations on, 59–65, 71, 73; innovations in Pennsylvania's anthracite coal operations, 53–54; mapmaking in the curriculum of mining schools, 48–49; map scales, 29; the power to control mining and, 9, 10–11, 17–18, 221–22; precursors of, 18–19; in the preindustrial era, 17, 37–40; preservation of the maps of closing or abandoned mines, 222–27; professional identity of mining engineers and, 49–52; storage of, 27, 33–35; surveying underground (*see* underground surveying); use of symbols in, 28–29. *See also* anthracite coal maps/mapping; Drifton Colliery maps
underground mining: dependence on technology, 7–8; development in America, 6–7; mines as technological systems, 8. *See also* Anaconda Copper Mining Company; anthracite coal mining; industrial mining
underground surveying: in the